T0231542

Competitive Innovation

Innovation

and

Improvement

Statistical Design and Control

Competitive
Innovation
and
Improvement

Statistical Design and Control

Kieron Dey

CRC Press
Taylor & Francis Group
Boca Raton London New York

CRC Press is an imprint of the
Taylor & Francis Group, an **informa** business

A PRODUCTIVITY PRESS BOOK

CRC Press
Taylor & Francis Group
6000 Broken Sound Parkway NW, Suite 300
Boca Raton, FL 33487-2742

© 2015 by Kieron Dey
CRC Press is an imprint of Taylor & Francis Group, an Informa business

No claim to original U.S. Government works

Printed on acid-free paper
Version Date: 20140729

International Standard Book Number-13: 978-1-4822-3343-8 (Hardback)

Visit the Taylor & Francis Web site at
http://www.taylorandfrancis.com

and the CRC Press Web site at
http://www.crcpress.com

Contents

Preface

This book combines two widely known statistical methods in a novel way to solve business, government, and research problems quickly, with sustained results. Illustrated by real large-scale case studies from a diverse range of industries and problem types, a simple unforgiving test is always used: sudden sustained improvement (or innovation) occurs. This is a high standard but one any organization can master with existing resources and staff, releasing latent energy rather than adding work for busy people. Innovation and improvement by design are explained, which opens up left-brain analytics to right-brain creativity and therefore to more people.

The two methods are statistical design (also known as experimental or orthogonal design) and statistical control (originally called economic control). Their flexibility solves large-scale problems simply, no matter how complex. Because the problem-solving strategy employed is a purely scientific method, integration into any existing problem-solving or research methodology (e.g., Six Sigma, comparative effectiveness research) is easy. The scientific method is shown within everyone's capability, in fact it is innate.

The book is written so that anyone can read and use it, including executives, managers, statisticians, scientists, engineers, researchers, and all of their supervisors and employees. Optional footnotes provide more advanced technical insight.

The cases show how to apply these methods to any business process, including those composed mostly of people (e.g., clinical care, educational performance, software design, website/search engine optimization, new product design/marketing, creative design/artistic layout, all sales channels including "feet-on-the-street," cross-channel optimization, call centers).

No mathematical notation is used. Mathematics (as advanced as needed to solve the problem) is used throughout the text, but translated into little more than simple arithmetic. Mathematics is perhaps to problem-solving as toes are to walking. (They're essential but there's far more to it.) This book explains the rest of scientific improvement, especially how to manage it. The cases suggest that the real world (rather than the mathematics alone), reveals how things work and how to make them work better.

Minimal important references are provided, limited to a few essentials. Several are landmark texts and papers that will continue to provide new knowledge as experience is gained over the years. These few select references credit original technical contributions.

Acknowledgments

Thanks are due to the late Dr. George Box, FRS, one of the most important scientists of these (or any) times, for his generous, often remarkable insights over the years, for first suggesting the book be written and reviewing early drafts in 2011, and for the vision that it contain no mathematical notation (already published extensively) but include instead management and scientific aspects.

I also thank Ed Mueller, a veteran CEO, who saw the utility of statistical design and control on large-scale problems, explained how to integrate them into management and pointed out that the science should include the organization.

For countless hours of technical discussion over the years I thank: Tim Baer, Mark Black, Randy Brown, Dave Coit, Dave Courtney, Brad Carlin, Steve Flint, Ron Fritz, David Futrell, Mike Gallagher, Steve Grady, Brian Joiner, Joan Keen, Dave Martin, KK Moore, Blaine Nelson, Ron Osminkowski, Chuck Proctor, Jorge Romeu, Bob Smith, Ian Worden and in management: Eddie Black, Debi Cahall, Iain Calder, Michael Canon, Ed Carroll, Stephen Carter, Randy Conway, Chuck Feltz, Peter Hager, Lee Hord, Neil Ismert, Mike Kaufman, Ted Loken, Charlie Roesslein, Luann Widener and Karen Wohkittel; also hundreds more executives, managers and employees who brought the field work to life.

For valued expertise, support and creativity, my thanks to Editor Michael Sinocchi and the team at Taylor & Francis: Rob Calver, Josie Banks-Kyle, Amber Donley and Iris Fahrer.

K.A.D.

1

Simplicity of Statistical Design and Control

Good simple ideas ... are our most precious intellectual commodity, so there is no need to apologize for the easy mathematical level.

Bradley Efron[*]

1.1 MAKING A START

Statistical design [1] and control [2], which find business solutions by testing then managing their implementation, are fundamentally simple. It's the need that's more important than the method though, such as improving health for thousands of people, or innovation of a new product in retail with strong sales maintained from the outset, such as shown in this book's opening, real cases. Statistical design and control is a management tool, not one that can be deployed by technical people alone. It's not top-down but starts and stays there. The cases show how and why.

We learn to talk by trying to, and then learn the rules later. It would not work the other way around. Statistical design and control is much the same way, except the craft is not as fully defined by rules (of mathematics rather than language). This is unusual (or at least far more pronounced) among analytical techniques. Like conversation, problem-solving is improvised, using rigorous rules. This is because each problem is unique (just like all the rest).

---

[*] Efron, B. (1982). *The jackknife, the bootstrap, and other resampling plans.* Society of Industrial and Applied Mathematics CBMS-NSF Monographs, 38.

No matter how skilled the technical leader, the only time to run statistical design and control perfectly is after it just finished. It's innovative in uncharted waters, thus it is impossible to know everything later known with hindsight. This doesn't mean poor designs are just fine; it means good design is imperfect.

The text is interspersed with exercises (mainly for technical people but most can be answered by anyone) with answers implicit in later text or in a later chapter. The full answer set is provided in the appendix. These exercises try to simulate some of what is gained by experience. The methodology is one of those that's easier to learn and master hands-on than only read about. It's self-teaching when used.

Problem-solving for improvement of existing processes, and innovation in the first place, both use the same approach (inasmuch as both seek solutions that work suddenly and can be sustained long term).

Throughout the book, formal definitions of statistical terms now in common usage are not given. Instead, enough for the layperson to get by is provided, appealing also to common sense and context. Colloquial terms in everyday usage (e.g., bell-shaped curve) are used instead of more formal terms (e.g., normal or Gaussian distribution). This is for simplicity and to keep the forest of practical improvement/innovation clearer than its trees.

Translations found useful to nontechnical people when doing statistical work in the field are provided in shaded boxes, placed in the text just before first needed.

Calculation methods are provided in shaded boxes also to set them off, again at first usage. Calculations using these methods in the cases are also in shaded boxes. These can be skimmed or skipped if preferred without losing the case stories or the text's tenor. Excess decimal places are used where they clarify the calculation method.

1.2 HOW DOES IT WORK?

The idea of statistical design is to find out how something works, so that it can be made to work better. It establishes cause and effect directly. Typical examples of outcome measurements to improve (and where) include:

- Health (nurses providing clinical care)
- Retail sales (stores)

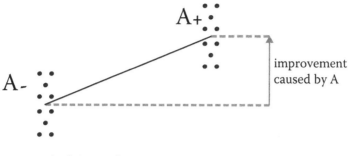

• test units (10 vs. 10)

FIGURE 1.1
Testing against a control.

Traditionally, a potential solution (labeled A for now) is tested (+) against a control (–); see Figure 1.1. The *test units* (e.g., nurses or stores) are randomized to test or control. The outcome is implied by the vertical axis where up is good (e.g., health, sales). Everything works the opposite way where down is good (e.g., hospitalizations, returns). This test design appeals to our intuition in being analogous to a balance scale to compare two weights.

Statistical design can test two potential solutions (from hereon called *interventions*) labeled A and B for now, without larger sample size or resources; see Figure 1.2. The same 20 test units are used as for A alone. Again the test units are randomly allocated.

This also appeals to our intuition, being analogous to a small aircraft with only a few occupied seats, where passengers will be moved to the front or back and left or right, to balance in both directions for flight. The imbalance in Figure 1.2 reveals both solutions help. The right angle shown (∟) ensures neither solution interferes with the other being measured. This is more easily seen by comparing back to Figure 1.1 in which the amount of improvement caused by A helping is unchanged by B also helping, in Figure 1.2. Instead of "at right angles," statisticians use the word orthogonality.*

---

* Technical people may prefer a little more formality. The inner-product of **A** and **B** taken as vectors of –1 and +1 is + 1 – 1 – 1 + 1 = 0 therefore since $\cos^{-1}(0) = 90°$, A and B are orthogonal, meaning "at right angles." That right angle is noted in Figure 1.2; the tests for A and B are called *orthogonal contrasts*. Because of this, (or just by analogy, that forces at right angles have no moment in each other's direction), the two interventions are estimated independently. An easier way to see this is to run correlations on the ±1 matrix in which the correlation of A with B will be found to be zero. The interaction also turns out to be orthogonal to A and B, as can be checked by augmenting the design matrix with AB (formed by multiplying A and B's elements into a new column AB: + – – + vertically), and using either method.

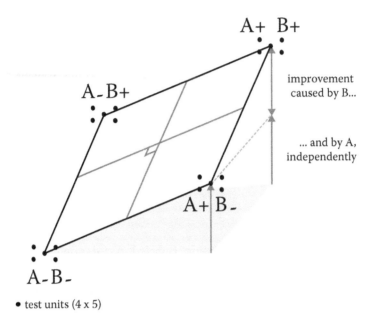

FIGURE 1.2
Statistical design for testing two solutions.

Because each solution is tested against its own absence, the term *counterfactual* is now used (instead of "control"). No attempt is made to hold other conditions constant. Instead, the real world (meaning an industrial or business process, existing or under development) is taken as it is found.

Using "–" for counterfactual and "+" for test, Figure 1.1 can be translated to:

A

–

+

and Figure 1.2 to:

A	B
–	–
+	–
–	+
+	+

Notice that this second test will also find out what happens when doing both interventions together. That's called an *interaction*, revealing if there's either synergy or diminishing return when combining the two potential solutions.

This simple idea of statistical design scales up to more than two interventions, still without more test units. (Statisticians refer instead to *experimental units*; the distinction in terminology is made here because testing is in the live business, not experimental mode.)

It turns out that the 20 test units shown are enough to test up to 19 interventions [3]; see Figure 1.3. This is powerful because whereas Figure 1.1 tests two (2) options (take or leave A) and Figure 1.2 tests four ($2 \times 2 = 4$) options (A, B, both, or neither), Figure 1.3 tests 20 of about a half-million management options ($2^{19} = 2\times2\times2\times2\times2\times2\times2\times2\times2\times2\times2\times2\times2\times2\times2\times2\times2\times2\times2$ = 524,288) to find the best, approximately.

The simple idea that two interventions were kept independent by the right angle shown in two dimensions (A and B) in Figure 1.2 scales up to 19 dimensions (A through S) where all pairs of columns are orthogonal, therefore they're all tested and measured independently. Statistical control is used beforehand to make sure the test units are similar enough in the

Row	A	B	C	D	E	F	G	H	I	J	K	L	M	N	O	P	Q	R	S
1	+	+	-	-	+	+	+	+	-	+	-	+	-	-	-	-	+	+	-
2	+	-	-	+	+	+	+	-	+	-	+	-	-	-	-	+	+	-	+
3	-	-	+	+	+	+	-	+	-	+	-	-	-	-	+	+	-	+	+
4	-	+	+	+	+	-	+	-	+	-	-	-	-	+	+	-	+	+	-
5	+	+	+	+	-	+	-	+	-	-	-	-	+	+	-	+	+	-	-
6	+	+	+	-	+	-	+	-	-	-	-	+	+	-	+	+	-	-	+
7	+	+	-	+	-	+	-	-	-	-	+	+	-	+	+	-	-	+	+
8	+	-	+	-	+	-	-	-	-	+	+	-	+	+	-	-	+	+	+
9	-	+	-	+	-	-	-	-	+	+	-	+	+	-	-	+	+	+	+
10	+	-	+	-	-	-	-	+	+	-	+	+	-	-	+	+	+	+	-
11	-	+	-	-	-	-	+	+	-	+	+	-	-	+	+	+	+	-	+
12	+	-	-	-	-	+	+	-	+	+	-	-	+	+	+	+	-	+	-
13	-	-	-	-	+	+	-	+	+	-	-	+	+	+	+	-	+	-	+
14	-	-	-	+	+	-	+	+	-	-	+	+	+	+	-	+	-	+	-
15	-	-	+	+	-	+	+	-	-	+	+	+	+	-	+	-	+	-	-
16	-	+	+	-	+	+	-	-	+	+	+	+	-	+	-	+	-	-	-
17	+	+	-	+	+	-	-	+	+	+	+	-	+	-	+	-	-	-	-
18	+	-	+	+	-	-	+	+	+	+	-	+	-	+	-	-	-	-	+
19	-	+	+	-	-	+	+	+	+	-	+	-	+	-	-	-	-	+	+
20	-	-	-	-	-	-	-	-	-	-	-	-	-	-	-	-	-	-	-

FIGURE 1.3
Multifactorial design for testing 19 potential solutions.

first place to support viable testing. Later statistical control is also used to manage, troubleshoot, and track successful implementation (which would otherwise be extraordinarily difficult).

A recent healthcare case follows to show how this is done in practice on a large scale.

1.3 CARE MANAGEMENT CASE: IMPROVING HEALTH FOR THOUSANDS OF PEOPLE

A new care management program was being established based on past practice. Care was provided by 20 nurses operating telephonically, serving about 2,500 chronically ill Medicare members living at home across several states. Each nurse had, in the past, been allocated about 80 to 100 patients to work with every couple of weeks. Patients in these programs typically stay a few months until reaching a renewed level of improved health using clinical criteria, so the nurses' lists of patients change over time. The allocation of patients to nurses is usually not random but depends on a few management criteria such as time zone, conditions, and each nurse's existing workload.

The single measurement of hospitalizations captures the entire story more simply than the original work, which included several other health measurements. The tradition in the industry is to measure hospitalizations per thousand members each month and multiply it up by 12 to an equivalent year. Here, that measurement turned out to be about 1,200 at the outset looking back over the previous year, meaning patients were typically in the hospital a little more than once a year (1,200 per thousand = 1.2 per patient). Hospitalizations are also known as admits, which includes readmits for any reason.

The project was completely defined and scoped in rough, in one hour, working with clinical and operational managers. (No problem needs more time, no matter how large or complex.) Often, briefings on the business are offered but instead, several questions were asked here that led more directly to the heart of the problem. Briefings do not usually include the salient information needed to start improving something.

Questioning started with the objectives, measurements, and the ways in which patients are reached, with a view to improving their health. What the scientist does in the opening meeting is understand how the thing works and where the triggers are that can make it work better. A reasonable grasp of terminology is needed before starting work in order to communicate efficiently. There is little technical value to knowing those names of things, because they are not helpful to understanding how the thing works and sometimes obscure it. Improving something uses different knowledge than is used to manage it. So it's not necessary (and counterproductive) for the scientist to learn everything the other experts already know. Conversely, it is harder to improve something solely with the knowledge needed to manage it. So a useful collaboration occurs between the scientist and other experts when improving something. The meeting broke up with an agreement to meet again the next day with all nurses, staff, and medical directors.

At this early stage it's already clear why statistical design is a senior management tool but is not top-down. Resources and meetings needed management support and legitimacy but tomorrow the creativity would be voluntary.

1.4 DISCOVERY

The main point of discovery, starting at this working meeting, was to understand further how the thing worked and start listing tactics that might make it work better. The guiding question was: "What do we think might improve health for the members and thus reduce hospitalizations?" The discussion kept pulling back to this single question. It's hard for groups not to digress into explaining more detail than is yet needed, therefore a useful, repeated follow-up question is: "In seven words or less, what would you change?"

Brainstorming (or ideation) is a popular format and is useful for some management problems. This is not so with this exercise, because the scientist has to converse and probe to keep developing that understanding of how the thing works. Also, the scientist has to lead the group into a sound list of changes to try and this relies upon experience, with short examples from similar cases thrown in to stimulate the thinking. Quite often, one small voice suggests something that's not credible with anyone else. It's the scientist's job to seize that spark and amplify it to reveal the full original idea. If the scientist were not out in front on this, the group could pounce on it with derision and a minor melee could ensue. Other ideas will be divisive, with factions lining up for and against them. These controversial ideas tend to have a higher likelihood of helping. Of course it's important not to allow just any crackpot idea. The distinction is that the small voice or warring faction will have noticed something in day-to-day work and is putting it forward to see if it will always work.

The more educated the group (especially if in technical fields) the harder it is for them not to continually ask how each tactic could ever be tested. This can morph into showing why this idea or that one could never be tested in practice. Instead, it's important to concentrate on the list of tactics first. The test design will come next but not yet. If it's attempted in real-time this early, it will be too difficult or confusing.

Requests by technical people inexperienced in statistical design to sit in on these sessions to learn how to assemble a design will tend not to be useful. Watching this part is a little like watching a violinist tune and warm up before a concert. It's not easy to see what he's doing or why and it doesn't sound impressive. What is very useful is an apprenticeship, learning what is going on behind the scenes and in the back of the scientist's mind, in order that solo projects can be started soon. Part of that is indeed sitting in on one or a few of these meetings to absorb by osmosis how it's done. If so, each meeting needs a "postmortem" taking five to ten minutes to explain what just happened and how it is starting to play into the statistical design. These raw tactics are then turned into interventions for testing: by defining them rigorously and writing a short standard for each.

To maximize improvement, each intervention *level* (i.e., the change [+] relative to status quo [−]) should be set as boldly as possible, without breaking or risking anything. Paradoxically, seemingly subtle changes often have large effects and are important to include.

Common errors in forming interventions are

- Listing too few or whittling the list down to a handful, under various rationalizations.
- Rationalizing that an intervention would only touch (say) half (or 15%, etc.) of the cases/transactions, and thus should be dropped. Strong findings have occurred, often when the intervention does not work as people intended (and may even "touch" no one with hindsight, or rather the real touch was not as anticipated). Statistical design is to find out what happens when we interfere with something, rather than guessing what ought to happen.

- Perceiving "we're already doing that" (when the facts would find that less than fully so).
- Finding a proposed intervention not "credible" (whereas there is only a small correlation between what works and what's initially thought or established in the literature, etc.).
- Tweaking the original spark until it's stifled, stomped, or dead.
- Arriving at interventions by consensus (whereas many break-throughs are from one small voice, aberrant, contrary, or across the flow); often the more contentious elements are found to have large effects when tested.
- Missing the angle that interventions can be the removal of current standards and need not always be incremental or additive. This "untesting" also reduces work and frees up capacity for what really helps.
- Multitudes of interventions where no one remembers the physics and so on behind each element on a sterile list; instead a manageable list is written up during each research session, documented as needed in real-time and kept as a living document with backup.
- Declaring: "We don't need to test that! We just need to go ahead and do it." That can sound good until some of these apparently obvious changes hurt (learned from the experience of hundreds of cases) and if done in this way, the statistical design would not know it.

Experience with statistical designs finds that they almost invariably serve up surprising results. So too will the interventions therefore tend to be somewhat surprising to many people initially (often on the seemingly unimpressive end of the spectrum).

It is important that one statistical head stay with the interventions from inception to test to analysis to implementation management. That head has to know the history and pedigree at all times. It's fine to have more than one statistical head collaborating from the start, as long as only one is responsible for the design. (This is also a good way to apprentice in statistical design and control.)

From time to time, someone offers to list the interventions to save the scientist's time. This would make about as much sense as someone offering to get your flu shot because you're busy. Developing interventions, some of which later help, is demanding work, making heavy use of experience, rather invisibly to onlookers.

TABLE 1.1

Interventions

A:	Medications review using a creative way with patients unable to recall or read labels accurately.
B:	New symptom education and management given patient history.
C:	Educational kit customized to patient's disease(s).
D:	Counseling for caregiver.
E:	A new way to manage transportation pickups reliably, to and from the primary care physician (PCP).
F:	Letter to PCP outlining patient's condition and suggesting enhancements in care.
G:	Scripted financial discussion and problem-solving.
H:	Referral if a particular health measurement was high.
I:	Home visit by nurse.
J:	Discuss advanced directives.
K:	If unable to reach (UTR) by phone, send letter instead requesting patient to call in.
L:	Coordinate home healthcare providers if applicable.
M:	Social worker visit if certain criteria met.
N:	Nurse caseload/frequency of care calls (100 patients called every two weeks vs. 150 called every month).
O:	Corporate office vs. work from home.
P:	Letter to patient listing questions to ask their doctor (PCP).
Q:	Slightly fitter vs. slightly sicker patients.
R:	New transition call after discharge from hospital.
S:	Early warning system for exacerbations.

The meeting ended with a list of 19 tactics to try to turn into interventions in the coming days. A few people were asked to develop a simple procedure for each, within the week. Some potential interventions needed more statistical work. The intervention tactics fermented over that time, as the details were worked out.

The full list, labeled alphabetically to keep track, is provided in Table 1.1. That list of interventions was mostly changes incremental to the status quo clinical model, with a few trying two options (this vs. that): interventions N, O, and Q.

Exercise 1

Is it important to standardize completely each intervention? Is a little variation alright?

1.5 MEASUREMENT QUALITY

All measurements have error, meaning they fluctuate about truth (not necessarily that someone made a mistake). It's essential to quantify the degree of error and decide if it will support a viable statistical design. This uses rigorous rules that almost anyone can learn and apply.

Because claims data take a few months to mature, authorizations data were used. Authorization occurs typically before hospitalization, when the health insurance plan approves the impending expense. The claim happens after hospitalization and is used for several purposes, including invoicing the plan.

Throughout the industry, the perception at that time was that authorizations data could not be used because they can be incomplete or err widely from the eventual claim (which was regarded as the "gold standard"). Several analyses over the years suggest the industry is right, except in a randomized statistical design. The perception here (widely held and at all levels) was that authorizations data were captured perhaps 10–30% of the time, underlining its unsuitability for this project.

A designed sample of claims and corresponding authorizations data took 15 minutes to find that about 90% of claims also had an authorization record. The reason for the pessimistic perception traced back to previous years' history when indeed that had been true and the data showed as much. The perception had not been updated although when a change was made in utilization management, without anyone noticing, the authorizations had become far more complete and reliable.

TRANSLATION 1

In field statistical work, translations found useful to nontechnical people have been:

Significant: A result large enough to have been expected no more than 1/20 by chance.

Correlation: The extent to which things vary together directly (e.g., walk speed and pulse) or inversely (e.g., distance driven and fuel remaining), though correlations may or may not be causal, generally.

Turning to how well hospitalizations correlated from the two record types, still in that 15-minute work session, the correlation was found highly significant (which isn't much of a bar to clear) and more important, the measurement error consumed a relatively small percentage of the total variation in hospitalizations. However, there were "spikes," where groups of authorizations could be very different from the associated claims data at the nurse aggregate level, or in daily aggregates, and this fluctuated over the months. Now, these spikes are a problem.

Short meetings with a couple of directors were tense initially. The perception of authorizations data being unusable was so deep that the finding that it was close to being usable was not credible. The notional reaction was that if the two records were close then the authorizations data must be wrong and that had to be fixed first. This idea was pretty close to what was needed statistically, so we listed why it was felt those data could be in error. Duplicates, nonhospitals (e.g., special needs facilities), and a few other things were suspected. Indeed we soon found several of each in the spikes, in both authorizations and claims data. Each spike was investigated until themes emerged leading to further changes to the data pull. Every large spike was understood; then the lion's share of the causes quickly corrected. This all took a couple of hours for one analyst. A few mild spikes remained. We were careful not to work on anything but the large spikes and to avoid any heroic efforts that could not quickly be copied routinely in future.

When the initial measurement work was finished, authorizations and claims data agreed even more closely. More challenges were proposed (inasmuch as it was felt this still could not be right) and the mild spikes were bored in on. A second round of detective work took a few minutes and even mild spikes were gone after some more programming, with the authorizations versus claims agreement still closer.

The perception of authorizations data being unusable began to crack and the statistics started to gain credibility. There was no hurry on this social front, because the statistical design would take awhile to run and we could worry about this nontechnical aspect later. There was, however, one more request: that, because we had made such rapid progress improving the measurement system, why not "perfect it" so that all records agreed perfectly. It turns out that once spikes are removed, any further effort to attain measurement perfection will be uneconomic and can worsen the measurement error. Managers listened to this and agreed to let the request for perfection go. The mathematics that provide this surprising insight were left until a later date because a full case review at completion would put it in better context. Along those lines of not explaining too much too soon, Chapters 2 and 3 provide the mathematics, using the actual data for this case.

This was the first of several examples in this book of exploding folklore. It was two months before the organization fully accepted the authorizations data provided they were used only for randomized statistical designs. By now, the measurement work had led to strong interest in using authorizations data, since it preceded the more slowly accruing claims by several weeks. Accordingly, mathematical models were developed (using data from both sources) to predict mature claims from the much earlier authorizations. These predictions saved up to four months. These did rather well in the months ahead and included a margin of error that was always found correct. Managers became familiar by accosting the scientist in the hallways each week to mention the actual versus prediction was up this much or down that much. Reminded of the margin for error (that always worked), full confidence was gradually established.

Measurement controls were put in place so we'd know the amount of measurement error each month and for each nurse. These answered all measurement error questions but preempted most. Chapter 4 provides those controls in full, with the data as they were used.

Still under the heading of measurement, the group of patients in a test is called a *cohort*. Because patients came for a few months and then went, according to clinical progress, the cohort was constantly shifting. That's called an open cohort. A closed cohort was used instead, meaning the fixed list of all patients in the new program just before the test started. This operationally defined the measurement whose quality was by now locked down. It would do no good to have quality measurement if the list of which patients to measure were left undefined. This has to be done early so it's entirely objective. Accordingly, new patients entering the program after its start were still in the test, but their data were excluded. This closed cohort gradually wears out over time but was found to decline only very slightly over a few months, which was all we needed.

1.6 CARE MANAGEMENT STATISTICAL DESIGN

As in Figure 1.3 earlier, the statistical design was prepared; see Figure 1.4. The randomization [4,5] of nurses to the design (using the original order of nurses listed alphabetically) is included. Each nurse would test each

Nurse	A	B	C	D	E	F	G	H	I	J	K	L	M	N	O	P	Q	R	S
2	+	+	-	-	+	+	+	+	-	+	-	+	-	-	-	-	+	+	-
7	+	-	-	+	+	+	+	-	+	-	+	-	-	-	-	+	+	-	+
18	-	-	+	+	+	+	-	+	-	+	-	-	-	-	+	+	-	+	+
1	-	+	+	+	+	-	+	-	+	-	-	-	-	+	+	-	+	+	-
5	+	+	+	+	-	+	-	+	-	-	-	-	+	+	-	+	+	-	-
19	+	+	+	-	+	-	+	-	-	-	-	+	+	-	+	+	-	-	+
9	+	+	-	+	-	+	-	-	-	-	+	+	-	+	+	-	-	+	+
16	+	-	+	-	+	-	-	-	-	+	+	-	+	+	-	-	+	+	+
14	-	+	-	+	-	-	-	-	+	+	-	+	+	-	-	+	+	+	+
4	+	-	+	-	-	-	-	+	+	-	+	+	-	-	+	+	+	+	-
8	-	+	-	-	-	-	+	+	-	+	+	-	-	+	+	+	+	-	+
3	+	-	-	-	-	+	+	-	+	+	-	-	+	+	+	+	-	+	-
13	-	-	-	-	+	+	-	+	+	-	-	+	+	+	+	-	+	-	+
11	-	-	-	+	+	-	+	+	-	-	+	+	+	+	-	+	-	+	-
17	-	-	+	+	-	+	+	-	-	+	+	+	+	-	+	-	+	-	-
20	-	+	+	-	+	+	-	-	+	+	+	+	-	+	-	+	-	-	-
12	+	+	-	+	+	-	-	+	+	+	+	-	+	-	+	-	-	-	-
15	+	-	+	+	-	-	+	+	+	+	-	+	-	+	-	-	-	-	+
10	-	+	+	-	-	+	+	+	+	-	+	-	+	-	-	-	-	+	+
6	-	-	-	-	-	-	-	-	-	-	-	-	-	-	-	-	-	-	-

FIGURE 1.4

Randomized care management statistical design for testing 19 interventions.

intervention having a "+" in her row. Notice each nurse tested about half of the interventions except in the last row where nothing is tested. This was popular as some nurses had wondered if they'd have to test all 19. In some cases that's preferred and the signs can then be reversed.

Exercise 2

What is the purpose of randomization?

At first sight it may seem we're asking a lot to test 19 interventions with just 20 nurses. In fact, because the design is orthogonal, we're really testing each intervention with 10 nurses against 10 (about 1,250 vs. 1,250 patients) and all interventions are kept independent, just as in Figure 1.2 except now instead of two dimensions there are 19. It's easy to tame the 19 dimensions, however, and see Figure 1.2 everywhere: a quick look at any column finds 10+ and 10−. A quick look at any other column finds 5+ versus 5− within the former 10+. So when any intervention is analyzed, all the others cancel out in the calculation. If an intervention is later found to help, by that calculation, it helps in the presence of 18 other variables also changing, rather than holding them constant. This previews an enormous advantage of statistical design in reflecting the real world, not an artificially controlled one. It makes the findings far more likely to be repeatable, without literally repeating the test. This is known as a *wider inductive basis* [1] (meaning 18 interventions "wider" or closer to the real world; *inductive basis* meaning the foundation for generalizing the test's findings at implementation).

Although the design is equipped to test 19 interventions, in practice only a few will be significant. In that sense we're really only testing a handful of interventions that will be revealed with test hindsight. This is called *effects sparsity* [6].

Exercise 3

False alarm means in this context, crudely speaking, a significant intervention result "due to pure chance" (whatever that really means). It might appear the false alarm rate will increase with the more interventions tested. Why does it in fact decrease? Why does that pose a larger false alarm rate problem for smaller designs?

1.7 BASELINE DATA

It's reasonable to question whether, inasmuch as nurses will differ in their hospitalization rates (largely because of the differing patients they each have, not being randomized), the impending test will be contaminated. In fact, it's not the differing that matters but the overall pattern in those differences among nurses. If there are large spikes in admit rates for particular nurses, then that will disrupt the statistical design, therefore in that case there are easy workarounds.

The average hospitalization rate in Figure 1.5's actual data from the month before the test means patients are typically hospitalized about once a year ($\sim 1,100/1,000 = 1.1$). No single plotted point differs markedly from the others, so nurse admit rates are homogeneous.

Nurses and supervisors found homogeneity easier to think of as a "level playing field" (as suggested by Figure 1.5). The analogy was helpful

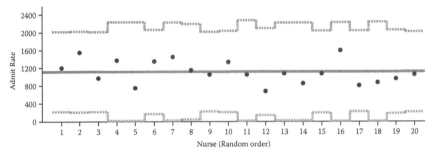

The dotted lines are 3 standard deviations either side of the average, and are tighter for nurses with about 150 patients vs.100. [u-chart, Figure 1.6 "pre"].

FIGURE 1.5

Hospitalization rate by nurse: homogeneity check using one month's data pretest.

because in sports the playing field is not level like a sheet of glass, but has little bumps and imperfections. All we care about is that there be no large molehills or potholes that someone could trip over or break an ankle in and that there is no serious slope that could disadvantage one team. It's the same thing when preparing for statistical design.

The dotted lines on the homogeneity chart provide a yardstick; when crossed, a nurse's admit rate would differ from the rest. Interestingly

TRANSLATION 2

Variation (variability): Fluctuations in an outcome.

Average: Always the arithmetic mean in this book, written as $\bar{y}$ (pronounced "y bar") for a variable y and calculated by adding up all the data in a sample then dividing by sample size.

Standard deviation: "Typical" deviation of single data points from the average. Written as the Greek symbol σ (pronounced "sigma") and estimated from data as $\hat{\sigma}$ (pronounced "sigma-hat") where the $\hat{}$ means "estimated from data." Different calculation methods than in most texts and courses in statistics are used in statistical design and control. They are provided just before needed.

Variance: σ^2.

Noise: Engineer's term for random variability and used later in significance calculations.

For a pure bell-shaped curve, 99.73% of data would fall within $3\hat{\sigma}$ either side of the average.

enough, there is no such thing as nurses being "too different" in their admit rates for a viable statistical design. They can vary as much as they like. If they vary a lot (whatever "a lot" means) then this will flush out when the calculations are done to determine how many nurses are needed. The solution to nurses differing "a lot" would be to test with "a lot of nurses"; up to 75 have been used.

1.8 MANAGING THE TEST

Once each intervention was written into a procedure, the 20 rows of Figure 1.4 were produced as 20 packages. Stacks of individual procedures were thrown on a table and a little assembly line stapled the 20 packages together just by following the "+"s, row by row. Then, the randomly allocated nurse's name was written on the top and the final stack of 20 packages readied for a meeting to explain everyone's role, answer questions, and give out the packages.

The planned duration of the test was three months. The test began and was managed with light-touch adherence monitoring and feedback. N, O, and Q were managed centrally so the nurses didn't have to worry about those. Nurses therefore typically had between 7 and 10 interventions (by test row) that they had to manage, with the last row null.

Nurses found their sublists of interventions challenging at first. After a week or so they had come down their learning curves and mastered the test. This is typical, with a test involving simple work usually needing about three days to get used to without disrupting the live business, or for more complex work such as nursing, a week or two. That light-touch adherence plan was critical, as opposed to a rigidly enforced test following test procedures 100.0%. All we really needed to know (early and continuing) was whether each intervention was being followed most, half, or some of the time, for corrective action.

This approach to testing has an element of laissez-faire, rather than finding out the effects of rigidly enforced interventions. The idea is to test more realistically net of imperfections and management realities. In this way, each intervention is measured in the real world of a live business, not under a facade of pristine perfection that could not be managed long term. If such a Pyrrhic test were attempted, the amount of improvement calculated at its end would be optimistic and would disappoint when implementation fell short later.

This approach is called *intent to treat*, because the intention is to manage all interventions well, while also letting nature take its course. Intent to treat is also important to avoid risk. In the instructions given to the nurses by their management, professional skill and judgment were, as always, given priority. If specific interventions were not suited to particular patients then the nurses had the freedom to make the decision.

Occasionally, with the best of intentions, an intervention that looked good on paper will be found flawed when tried. It will then be found slight on adherence and formally dropped after a few days. This is quite rare but does need to be a built-in safety valve. None of this (no matter what transpires) contaminates the test, it being orthogonal. If we meant to test "this" but it morphed to "that," then with hindsight, the actual intervention measures the effect of "this and that" (even if dropped). A week of "this" then "that" for the remaining test isn't a problem either, inasmuch as the design is orthogonal.

Exercise 4

Is intent to treat absolute or a question of degree? Would an intervention adhered to half the time be acceptable? What about 20% or −30%? Does each answer change with hindsight if the test later finds the intervention significant?

Over the years (especially in professions such as nursing where strict protocols are well followed) the adherence has been measured as high as 95% to 100%. Whether to manage lower adherence higher, or let it stay at its own level, or stop the intervention (rarely) depends on the details. It's not necessary to know the exact percentage of adherence by intervention. No one would know what to do with that anyway. It just needs to be known whether each intervention is being done most, half, or some of the time, for management purposes. Anything more will wear people out with administration.

This practical issue, that some interventions will not apply all of the time or may not work well, leads to a vital part of the scientist's broader work. She has to know what went on at all times and places in the test. This is just a continuation of the earlier point that the scientific work is to understand how the thing works and how to make it work better. In the test, that intensifies: partly to advise on flawed interventions and (usually more important) to know what was behind the analysis, after the test. Statistical design cannot be done locked up in an office pumping out the mathematics and so on. The scientist has to be out in the trenches, chatting informally around the coffee pot, standing behind the test design, and making it work.

Professional statistical duties are often those of a neutral independent observer. Statistical design is different. The scientist has to be interferer-in-chief (while remaining strictly objective in analyzing the data and using the findings). Not much can go wrong here though: either something improves by the calculated amount or it does not. The scientist's objectivity and analysis have to pass that unforgiving test just a few days after the technical work is completed.

Live business tests have to be deliberate in knowing what went on at all times and places. It's impossible to be everywhere at once, however, so the eyes and ears on the ground are helpful: those belonged to the nurses.

1.9 TEST RESULTS

The hospitalization rates for the test are reproduced in Figure 1.6. Recapping: the industry tradition is to measure the hospitalizations each month, divide by the patient count, then multiply by 12,000 to give equivalent hospitalizations per thousand patients per year. At implementation we modified this to weekly to gain more immediate feedback. The analysis could be conducted in an hour with pencil, paper, and Figure 1.6, just averaging test data for + versus −, for each intervention, before turning that page, or, to make sure the method is understood, just trying A. It's also vital to run the same arithmetic on the pretest data to give the natural inherent pretest biases.

1.10 EXPLORATORY ANALYSIS

Nothing more than simple arithmetic is used in this initial exploratory analysis. The story will not change with later assessment of statistical significance. Figure 1.7 gives the intervention effects on the hospitalization rate. A

Row	A	B	C	D	E	F	G	H	I	J	K	L	M	N	O	P	Q	R	S	Pre	Test	Patients	HCC*
1	+	+	-	-	+	+	+	+	-	+	-	+	-	-	-	-	+	+	-	1216.2	1146.0	148	4.1
2	+	-	-	+	+	+	+	-	+	-	+	-	-	-	-	+	+	-	+	1551.0	1357.8	147	3.9
3	-	-	+	+	+	+	-	+	-	+	-	-	-	-	+	+	-	+	+	973.0	926.3	148	3.5
4	-	+	+	+	+	-	+	-	+	-	-	-	-	+	+	-	+	+	-	1375.0	1504.5	96	3.7
5	+	+	+	+	-	+	-	+	-	-	-	-	+	+	-	+	+	-	-	750.0	1361.4	96	3.7
6	+	+	+	-	+	-	+	-	-	-	-	+	+	-	+	+	-	-	+	1353.4	1265.2	133	3.4
7	+	+	-	+	-	+	-	-	-	-	+	+	-	+	+	-	-	+	+	1469.4	954.2	98	3.4
8	+	-	+	-	+	-	-	-	-	+	+	-	+	+	-	-	+	+	+	1153.9	861.9	104	4.0
9	-	+	-	+	-	-	-	-	+	+	-	+	+	-	-	+	+	+	+	1040.0	1455.1	150	4.0
10	+	-	+	-	-	-	-	+	+	-	+	+	-	-	+	+	+	+	-	1333.3	1317.2	144	3.9
11	-	+	-	-	-	-	+	+	-	+	+	-	-	+	+	+	+	-	+	1066.7	1623.5	90	4.1
12	+		-	-	-	+	+	-	+	+	-	-	+	+	+	+	-	+	-	672.0	1057.4	125	3.4
13	-	-	-	-	+	+	-	+	+	-	-	+	+	+	+	-	+	-	+	1090.9	1747.4	99	3.9
14	-	-	-	+	+	-	+	+	-	-	+	+	+	+	-	+	-	+	-	848.5	899.3	99	3.3
15	-	-	+	+	-	+	+	-	-	+	+	+	+	-	+	-	+	-	-	1098.6	1605.7	142	3.9
16	-	+	+	-	+	+	-	-	+	+	+	+	-	+	-	+	-	-	-	1608.3	1193.2	97	3.1
17	+	+	-	+	+	-	-	+	+	+	+	-	+	-	+	-	-	-	-	800.0	1027.3	150	3.4
18	+	-	+	+	-	-	+	+	+	+	-	+	-	+	-	-	-	-	+	884.2	1350.8	95	3.0
19	-	+	+	-	-	+	+	+	+	-	+	-	+	-	-	-	-	+	+	963.5	1274.1	137	3.4
20	-	-	-	-	+	+	-	-	-	+	-	-	-	-	-	-	-	-	-	1061.2	1477.5	147	3.4

* $\overline{HCC}$ is the average HCC per nurse. HCC is an index of patient health in regard to conditions: the higher the sicker. It's correlated with hospitalization rate. Randomization prevents it from affecting intervention effects measured. This can be checked by copying this page, snipping the 20 rows into little strips and dropping them in a hat, shaking it well to mix up (randomize) the strips then dealing them into two heaps. The average HCC of one heap will be similar to that of the other heap. Repeating the exercise using any intervention's + vs. – to make the split will find the difference in average HCC score hardly related to intervention effects, if at all. Yet HCC and hospitalization rate are correlated in the original member level data. (HCC is rounded here to ease calculations.)

FIGURE 1.6
Hospitalization rates and other nurse level data* from the care management test.

dry-run analysis appears also: these are the biases for the period just before the test. All numbers can be checked using a calculator and Figure 1.6 data, just averaging + versus – in each column and subtracting the latter from the former. The dry run uses pretest data; the test analysis uses the test data. Other than that, the dry run and test analyses are exactly the same calculations. Then follows an analysis of the change in hospitalization rates since the pretest. Figure 1.8 reproduces the interventions in abbreviated wording, making clear what each intervention is testing (+) and its counterfactual (–). This makes it easier to see what the results mean, physically.

It might seem comical on that dry run: analyzing the test before it runs. That's its whole point though: ideally it would give all zeros as the intervention effects before the test starts. In the real world, however, they

Row	A	B	C	D	E	F	G	H	I	J	K	L	M	N	O	P	Q	R	S	Pre	Test	Change	Patients	HCC
1	+	+	-	-	+	+	+	+	+	+	+	+	-	-	-	-	+	+	+	1216.2	1146.0	-70.2	148	4.054
2	+	-	-	+	+	+	+	+	-	-	+	+	+	+	-	-	-	-	-	1551.0	1357.8	-193.2	147	3.939
3	-	-	+	+	-	-	+	+	+	-	+	-	+	+	+	-	+	+	-	973.0	926.3	-46.7	148	3.493
4	+	+	+	-	-	+	+	-	-	+	+	+	-	+	+	+	-	+	+	1375.0	1504.5	129.5	96	3.741
5	+	+	-	+	+	-	+	+	-	+	-	+	+	-	+	+	+	-	+	750.0	1361.4	611.4	96	3.702
6	+	+	-	+	-	+	-	+	+	-	+	+	+	+	-	+	-	+	+	1353.4	1265.2	-88.2	133	3.359
7	+	+	+	-	+	+	-	-	+	+	-	-	+	+	+	-	+	-	+	1469.4	954.2	-515.2	98	3.399
8	+	-	+	-	+	-	-	+	+	+	+	-	-	+	+	+	+	+	+	1153.9	861.9	-292.0	104	3.971
9	+	+	-	+	-	+	+	-	+	+	-	+	+	-	+	+	+	+	-	1040.0	1455.1	415.1	150	4.020
10	+	+	+	-	+	+	+	-	+	+	+	-	+	+	-	-	-	+	+	1333.3	1317.2	-16.1	144	3.946
11	-	+	-	+	+	+	+	-	-	-	+	+	+	+	+	-	-	+	+	1066.7	1623.5	556.8	90	4.141
12	-	-	+	-	+	+	-	+	-	+	-	+	-	-	+	-	+	+	-	672.0	1057.4	385.4	125	3.376
13	+	-	+	+	-	+	+	+	+	-	+	+	-	+	-	-	+	-	+	1090.9	1747.4	656.5	99	3.923
14	-	+	+	+	+	-	-	+	-	+	+	-	+	-	-	+	-	+	-	848.5	899.3	50.8	99	3.276
15	-	+	+	-	+	+	+	-	+	+	-	+	+	+	-	+	+	-	+	1098.6	1605.7	507.1	142	3.943
16	-	+	+	+	+	-	+	+	+	+	+	-	-	-	+	+	-	-	+	1608.3	1193.2	-415.1	97	3.131
17	+	+	+	+	-	+	-	+	+	-	+	+	+	+	+	-	+	+	-	800.0	1027.3	227.3	150	3.399
18	+	-	+	-	-	+	+	+	-	+	-	+	+	-	-	-	-	+	-	884.2	1350.8	466.6	95	3.021
19	+	+	-	+	+	+	-	+	-	-	+	+	-	+	+	-	+	+	+	963.5	1274.1	310.6	137	3.395
20	+	-	-	-	-	-	+	+	+	-	+	-	+	+	-	+	-	-	+	1061.2	1477.5	416.3	147	3.359

Test:

	A	B	C	D	E	F	G	H	I	J	K	L	M	N	O	P	Q	R	S
Average +	1169.9	1280.4	1266.0	1244.2	1192.9	1262.3	1308.4	1267.3	1328.5	1224.7	1211.4	1293.4	1255.5	1255.4	1302.9	1245.6	1398.0	1139.6	1281.6
Average -	1370.6	1260.1	1274.6	1296.3	1347.7	1278.2	1232.1	1273.2	1212.1	1315.8	1329.1	1247.2	1285.1	1285.2	1237.7	1294.9	1142.5	1401.0	1259.0
Effect	-200.7	20.3	-8.5	-52.1	-154.8	-15.9	76.3	-5.9	116.4	-91.1	-117.7	46.2	-29.6	-29.9	65.2	-49.3	255.5	-261.4	22.7

Dry Run

	A	B	C	D	E	F	G	H	I	J	K	L	M	N	O	P	Q	R	S
Average +	1118.3	1164.2	1149.3	1079.0	1197.0	1139.3	1102.9	992.6	1131.8	1051.3	1189.3	1194.3	977.1	1091.9	1123.2	1119.6	1167.6	1104.5	1154.6
Average -	1112.6	1066.7	1081.6	1151.9	1033.9	1091.6	1128.0	1238.3	1099.1	1179.6	1041.6	1036.6	1253.8	1139.0	1107.7	1111.3	1063.3	1126.4	1076.3
Effect	5.8	97.6	67.7	-73.0	163.1	47.4	-25.1	-245.6	32.7	-128.3	147.7	157.7	-276.8	-47.1	15.5	8.3	104.2	-22.0	78.3

Change

	A	B	C	D	E	F	G	H	I	J	K	L	M	N	O	P	Q	R	S
Average +	51.6	116.2	116.7	165.3	-4.1	123.1	205.5	274.7	196.7	173.4	22.1	99.1	278.4	163.5	179.6	126.0	230.5	35.1	127.0
Average -	258.1	193.5	193.0	144.4	313.8	186.6	104.1	35.0	113.0	136.2	287.6	210.5	31.3	146.2	130.0	183.6	79.2	274.6	182.6
Effect	-206.5	-77.3	-76.3	20.9	-317.9	-63.6	101.4	239.7	83.6	37.2	-265.4	-111.4	247.1	17.3	49.6	-57.6	151.3	-239.4	-55.6

FIGURE 1.7
Analysis using only simple arithmetic.

	Counterfactual (−)	Test (+)
A: Medications review, more creatively	No	Yes
B: New symptom education and management	No	Yes
C: Educational kit customized to patient	No	Yes
D: Counseling for caregiver	No	Yes
E: A new way to manage transportation	No	Yes
F: Letter to doctor (PCP) to enhance care	No	Yes
G: Scripted financial discussion	No	Yes
H: Referral if a health measurement was high	No	Yes
I: Home visit by nurse	No	Yes
J: Discuss advanced directives	No	Yes
K: If unable to reach (UTR), send letter	No	Yes
L: Coordinate healthcare providers if applicable	No	Yes
M: Social worker visit if certain criteria met	No	Yes
N: Nurse caseload/frequency of care calls	150/month	100/2 weeks
O: Nurse office	Home	Corporate
P: Letter to patient with questions to ask PCP	No	Yes
Q: Patient health (by HCC score cutoff)	Fitter	Sicker
R: Transition call after discharge from hospital	Old	New
S: Early warning system for exacerbations	No	Yes

FIGURE 1.8
Recap of interventions.

average about zero and spread out, mostly as a roughly bell-shaped curve. Occasionally, pretest bias will be significant. Thus it makes sense to make sure none of these are driving the test results.

An early objection many reasonable people have (already defused in the homogeneity check) is that nurses will differ in their hospitalization rates (largely due to their differing groups of patients) so we cannot now have it both ways and ignore it. The earlier homogeneity check was an important prerequisite for a viable test. That will still leave pretest biases, however, hence this dry run.

Exercise 5

If someone said that both pretest and test were just picking up false alarms, all "due to chance," what would be the correct response?

It's common sense in Figure 1.7, using just the eye to glance along the top list of intervention test effects, to compare them to the dry run below it, and check off which interventions really help. Clarifying examples are A (which is almost unaffected by the dry run) and E (which overcame a dry run "headwind" and thus may be stronger than the test per se suggests). H, K, and M are especially interesting.

The change analysis is not ideal because it adds noise (by subtracting pre from test) but it's been found reassuring to nontechnical people to see this. It tries to do in one step what the test analysis and comparison to the dry run do. It gives a sense of leveling the pretest playing field by comparing nurses in the test to themselves beforehand.

1.11 WHAT MIGHT THE RESULTS MEAN?

It's important to talk to a few people to understand what the initial results might mean. This is done before presenting results, because if that were done prematurely we'd get rationalizations, and instead we need to get closer to explanations.

The largest effects at face value were obvious and were scrutinized. A and E were new, creative ways to manage two issues known to be important: medications review and transport to get Medicare members to their primary care physician for routine appointments.

N had been controversial when first suggested. It had been felt that nurses could not handle more than about 100 members each. The test tried 150 by making less-frequent calls, with virtually no effect on hospitalization rate. It has among the smallest effects in the whole list of interventions, including after taking pretest bias into account. Therefore the productivity can be increased 50%, help far more patients and reduce the frequency of phone calls. Everyone liked this and gained from it. It doesn't reduce hospitalizations per se but it will, by helping many more people for whom there would otherwise not have been capacity.

The result for O means nurses can work from home and free up office space in a growing firm. If there are other reasons to keep employees in a central office then no harm will be done. The effect is quite flat.

Q was designed to see if slightly sicker patients (i.e., higher HCC) would respond more. Of course it shows Q+ "hurts" (because Q+ selected sicker people). Comparison to the dry run reveals that selecting slightly sicker members than in previous experience does not prevent hospitalizations.

R encompassed a new way to work with the patient after discharge from the hospital and was the most effective intervention at face value. Nurses filled in why this was felt to be effective and what they had experienced when working with patients using it. They explained that this was well known to be important throughout the profession and there were standards on this. This particular way was found especially effective.

There were also surprises in some of the things that didn't help. Requests to break the data out to more specific patient subsets (e.g., diabetes sufferers only) made no difference. Once a statistical design has spoken, it's hard to get it to change its mind.

With the inner working of the test better understood, the analysis was completed and readied to discuss with the full group more formally. It's their thing (not the scientist's) therefore no PowerPoint was prepared with which to talk at them. The scientist's role is to catalyze the meeting and clarify what the statistics mean, supporting the management lead. The outcome is a correct implementation plan that will cause hospitalizations to be reduced, suddenly and sustained. It might be tempting for a statistician to present the results and pronounce judgment on what works and what doesn't with an already baked in crisp implementation plan. With statistical design, the odds are high (because the findings will usually be surprising) that someone will interrupt with, "Wait! Do you think that can be right? That makes no sense," which is the right question to the wrong person. Homework should have been done prior and those questions answered mostly informally by clinical and operations experts and workers, with a little research to solve the remaining puzzles. The meeting is to finish work and implement, not to start work without knowing what the data mean.

1.12 FINDINGS ARE OFTEN SURPRISING

About 30 clinical people had been asked to rank interventions before the test ended by voting for each one expected to help. The number of votes cast for each were

11: G
10: A
9: S
8: H
7: B, R
5: E, K
4: D, J, L
3: M, O, P
1: C, F
0: I, N

Interpretation is self-explanatory. The test took the collective to a place no one expected, and that the collective (20 nurses, plus staff, and managers) would not have otherwise found. Even A, voted almost top, had support from only about a third of professionals, and thus would have been unlikely for adoption. (Proof of this is easy: it hadn't been.) This of course doesn't detract from the clinical expertise used in identifying every intervention. Experience finds the best ideas come from experts plus frontline workers. Unfortunately it appears impossible, over hundreds of systems like this one, to rank correctly or close, in advance.

Until experiencing large statistical design, it would be natural to suppose the strongest helpful interventions would be near the top of the list. In fact, this case is the first of several that find helpful interventions are scattered from the top to the bottom of the list. They're impossible to pick in advance. Statistical design findings are usually obvious, with hindsight. They also are self-teaching in a rather natural way. Reasonable people might after awhile believe of the helpful findings that they could have told us so without the test. The proof that that is impossible is (a) they didn't, (b) to the extent the helpful findings were perhaps known by someone at some level, the more than a dozen things that did not help were equally "known" helpful and (c) the collective understanding is what matters in gaining support for an improvement. Statistical design aligns the collective voluntarily in a competitive direction previously avoided; see Figure 1.9.

A follow-up analysis (retrospective to include the three months of the test) a couple of months later found little change in the intervention effects except for P, which grew to −324.4 admits per thousand patients per year, becoming the largest effect. The nurses quickly realized why this was so. P was a letter prepared by the nurse for the patient either to give to her doctor

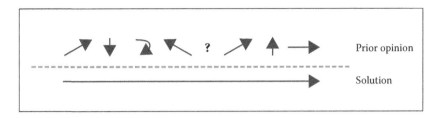

FIGURE 1.9
Statistical design aligns the collective voluntarily.

or read from it, putting important questions about her care. Apparently with this patient population there was often reticence about speaking up with doctors. The nurses explained it would have taken awhile for the letter to reach the patient, a doctor's appointment set and kept, and then the enhanced treatment to take effect. P had garnered just three votes prior.

1.13 SIGNIFICANCE OF THE RESULTS

Every intervention has a calculated effect but only the larger ones will bring real improvement over and above the variation first quantified in Figure 1.5. If no intervention had any effect, they'd average about zero and spread out like a bell-shaped curve. It's hard to guess, in Figure 1.10, which interventions really are large (either helping or hurting) although a few look promising. An ingenious device [7] is used to straighten bell-shaped curves into straight lines,* which the eye can see very well. Figure 1.11 re-plots the intervention effects in this way.

Significant effects tend to "peel off" to the top right (hurting) and bottom left (helping). If someone said, "Well, the whole thing looks like a line

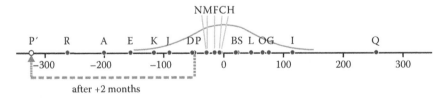

FIGURE 1.10
Intervention effects.

---

* The method is called the normal probability plot. There is also a half-normal plot that ignores the sign associated with each effect but advantages to the full normal plot will be found and used in later cases.

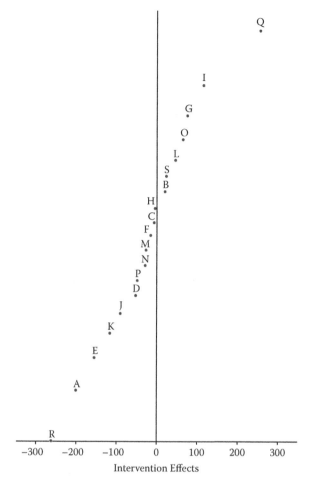

FIGURE 1.11
Normal plot of intervention effects.

P is from the original analysis, later growing to the largest
helpful effect at −324.

and it's not very straight," it's useful to hold the book at an angle so the
line of sight just glances the page. Then swivel the book so the line of sight
is also aligned with the plotted points. A clear straightish line appears
in its middle, with clear kinks on both ends. It also helps to fix the zero
(drawn as a solid vertical) as the center of the straight part. It's clear that
R, A, E, K, and perhaps J help; Q and perhaps I hurt.

The method might seem approximate. If nature worked as an on–off
switch, something exact would indeed be needed. Instead a practical

approximation is fine. Chapter 8 provides more sophisticated calculations of significance, but the story remains unchanged.

Life isn't as simple as finding the significant interventions, as if improvement were as clear-cut. Usually, in tests of this type involving people, some of the improvement occurs during the test. This is because of the standardizing effect of controlling so many interventions thought important. Interest therefore accrues in also implementing interventions that lean toward helping but are not strictly significant.

1.14 IMPLEMENTATION

The calculated improvement was about a quarter reduction in hospitalizations,* as a rough yardstick for implementation success. A+, E+, J+, K+, N−, O−, Q− and R+ were implemented initially and P+ followed within two months as soon as it was found. The implementation results are shown in Figure 1.12. The implementation started at week 6 and improvement began emerging especially in more detailed cohort analysis. P was implemented around week 15 as the strongest effect, but would have been expected to start showing a little later.

N was implemented with about half as many more patients added for each nurse. The total population in care was further increased by adding more nurses. This was quite a burden on the care management program as experienced nurses helped train new ones. The care program was much diluted for four to six weeks, really minimal. As it came back, the hospitalization rate visibly and steadily eased down. The reduction ended up with 15–20% improvement compared to the quarter reduction predicted from the test.

The finding that nurses could work with 150 members (not the 80–100 expected) allowed the quantity of members in care per nurse to be increased 50%. That and the extra nurses hired brought care capacity up to over 5,000 patients by week 18. The improvement held after the expansion.

---

* A quick calculation of the hospitalization rate expected at implementation is the test average (at 1270.3) plus half of each helpful effect. That half is because when calculating off the test average, each intervention was already half in play (half the nurses had tested it) therefore it cannot be double counted. Ignoring Q: the rough estimate using A, E, and R's effects is 1270.3 + (−200.7 − 154.8 − 261.4)/2 = 1270.3 − 308.5 = 961.8. Compared to the test average this is a calculated improvement of 308.5/1270.3 = 0.243 or about 24%.

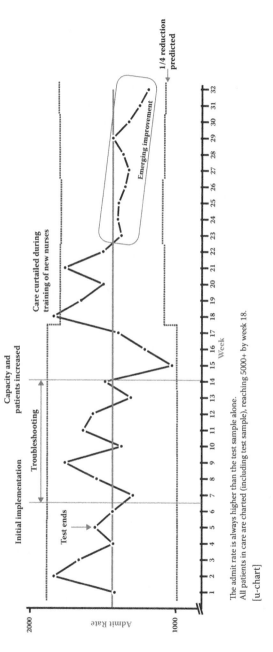

FIGURE 1.12
Implementation results.

The implementation chart (Figure 1.12) translates this narrative into a more objective statement. It shows all patients in care (including the test sample). The test sample experienced lower hospitalization rates, by design, than the untested patients, because the latter's HCC scores were higher, sometimes considerably so. Experience elsewhere had found that beyond a certain HCC score, patients did not respond as well to this level of care. The result for Q here leans to further proof. Accordingly, the implementation chart shows higher rates than the test sample. These details were tracked carefully in different cohorts, behind the scenes. Still, the practical emphasis had to be on the whole population.

The dotted lines are again just the average give or take three standard deviations, as used in the homogeneity check. The lines tighten as more patients are added. Ordinarily, for a bell-shaped curve, the dotted lines will be crossed on the low side 0.135% of the time (like a "14-year flood" for weekly data*) therefore Week 15 is noteworthy (just before the thousands of extra patients take it one step back again). After Week 23 the 10 points in a row below the average confirm real improvement to the nonstatistical eye (like 10 coin tosses landing heads with prior probability $(1/2)^{10}$ which is below one in a thousand). Notice also that week-to-week variation tightens at implementation. The two groups not in the test (Figures 1.13 and 1.14) give further reassurance but it's a weaker proof than the main chart data.

This gives the first glimpse of the single, simple, but unforgiving test of good statistical design and control: something improved suddenly then sustained. No such glimpses were available at the time, until a few weeks' hindsight, like the bottom of a recession. Therefore the critical early stages of improvement were done mostly in the dark, as follows.

1.15 IMPLEMENTATION TROUBLESHOOTING

Implementation doesn't happen as if by magic. A memo announcing what to implement will not work. Even in a process with static controls (such as a direct-mail piece or manufacturing) the organization will drift without real-time feedback control. There can be a perception that manufacturing implementation is easy inasmuch as the knobs can be set and left alone. That

---

* 0.135% = .00135, therefore the lower dotted line expects to be crossed naturally every (1/.00135) weeks = 740.7 weeks, 740.7/52 years = 14.2 years.

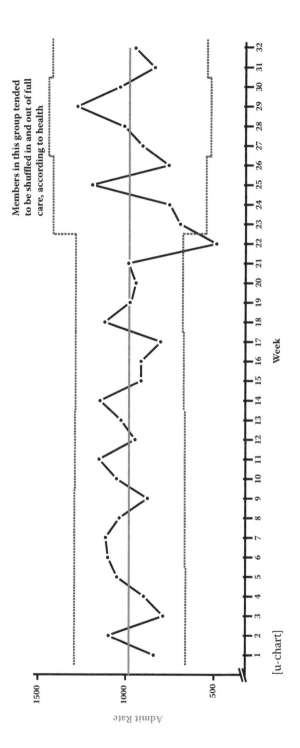

FIGURE 1.13
Similar fitter patients (limited care management).

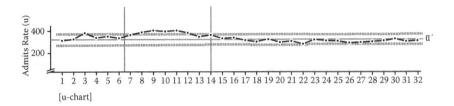

[u-chart]

FIGURE 1.14
Similar fittest patients (little or no care management).

would be true if fingers didn't twiddle knobs weighing years of experience more heavily than a recently competed statistical design, or if new fingers weren't ever hired. Direct-mail fingers are apt to twiddle in the same way.

Seedling improvement is a fragile thing preceding strong results. The scientist has work to do to ensure it's not pulled up to see how the roots are doing. Statistical control for this purpose is formalized in Chapter 3, but for now it's common sense. The device used to evaluate and control improvement is a seemingly simple chart (Figure 1.12). Anyone can read it with no mathematical knowledge and it becomes self-teaching with use. A little like reading music, it takes skill to read the dots initially, until people get the idea. Some have an eye for it, others don't, but all can pick it up rather fast. There's no room to misread the dots, as something will soon improve or not. Therefore there can't be any over- or under-interpretation: it has to be right on, as the rapidly approaching future improvement soon finds it.

Reasonable people tend to think improvement always should look like a cliff and start as clear as a Hollywood movie about science. Instead it's subtle at first and can be stomped dead far more easily. Not that the improvement is slight. Figure 1.12 shows a growing reduction in hospitalizations at $10,000 of cost each on average and perhaps a higher value than that for a health setback this serious for so many people. Every member tends to be in the hospital about once every 10 months in this chronically ill cohort. A few thousand members quickly adds up to a few million annually in cost alone.

At Week 14 the scientist determined improvement would fall short and met with the 20 nurses. The importance of knowing what went on at all times and places during and after the statistical design came to a head in this 30-minute meeting. It should be stressed that no one knew how the nurses would react, but all of the facts had to be ready, depending on which way the meeting went. Having listened to a random sample of 20 calls (one per nurse) the day before, the information that one of the helpful interventions was used by only four of the nurses was provided. The rough estimate of 20% adherence was therefore far short of full implementation.

Several nurses chipped in that the patients did not like that particular intervention, so it had seemed good service to drop it. This met with growing agreement from the group and had reasonable consensus. This is a common reaction and one usually overcome by managers pointing out that the objective is to reduce hospitalizations, not increase satisfaction in the fleeting moment of the call. Recalling the four nurses who had used the intervention, more detail was provided that all 4 calls had gone smoothly and the other 16 had contained no signal from the patients to avoid that particular intervention.

Because there was no evidence for the patients not liking the pharmacy review (and it had proved helpful in the test based on an original idea from the nurses themselves), the discussion continued. It turned out there was a simple productivity conflict that the nursing director immediately removed: the pharmacy intervention took several minutes. With the floodgates opened, the other interventions (averaging about 50% adherence) were explained more easily and those gaps closed too.

Usually this phase takes a couple of meetings a month apart. Here we just had the one. This facet of the case underlined why implementation is the hardest part (made straightforward in the next chapter by statistical control at full power) and why statistical design is a management tool aided and abetted by a technical expert. Notice that the implementation would not have happened if the manager had issued an edict for 100% implementation. This again underlines why improvement is not top-down but starts and stays there. Nurses had the freedom to take it or leave it. Had they not spoken up, we'd have been none the wiser and implementation would have failed or fallen short. Had the manager not solved the productivity clash, improvement would have been prevented.

In that sense this was always the nurses' and their managers' thing. Their ideas were all included in the testing. They explained why some things helped, once revealed statistically. They chose to answer the scientist's questions and changed the management system. There was no reporting line that required nurses to answer the technical questions, nor would one have helped. Instead, someone (or more than one person) in the meeting knew what had happened throughout the test.

Statistical designs have been found fun for everyone to do on serious business. Rather than adding work for busy people, they tend to release energy and squeeze more out of time when designed properly. They also unleash a natural skill people have and enjoy: solving problems. People find solving problems both fun and satisfying. If we didn't, we'd not do jigsaw puzzles, which have no other useful purpose.

The clean result in this case doesn't imply all cases go as smoothly. They invariably work but with varying degrees of teeth gnashing. In that this case was not special, it suggests that statistical design and control applies generally.

Exercise 6

What ways, other than Figure 1.6's footnote, might there be to check that the HCC score had no effect on the test findings under randomization? What statistical weaknesses would they have?

Exercise 7

Would a refining test to check that the helpful interventions really did help (and perhaps fine-tune them) have been advisable? Would a control group held out to confirm the implementation improvement be advisable (as well or instead)? Why or why not?

This completes the single unforgiving test that statistical design and control was done correctly: something improves suddenly then sustains. There would have been no mathematics to predict the final results, even from the earlier ones and all the data behind them. Conversely there are lots of ways, using mathematical models, to suggest the test's implementation could fail before it succeeds. All mathematical models are wrong to a degree, but the real world used here is always correct. Re-admits were also analyzed, with the helpful interventions all leaning to helping reduce re-admits as well, one significantly.

Dr. Randy Brown of Mathematica Policy Research Inc. in Princeton, New Jersey, conducted the independent validation of this case for publication, reproducing the results as above.

Figure 1.15 completes this opening case, with an illustration of intent to treat, used in all cases.

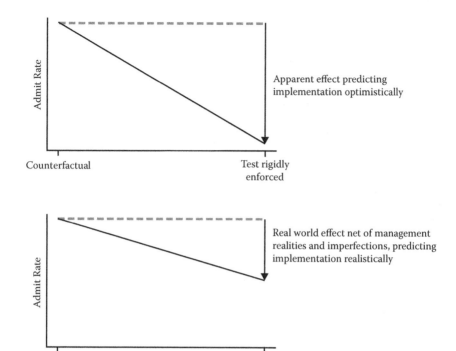

Intent to treat always applies, regardless of case or measurement.

FIGURE 1.15
Advantage of intent to treat in care management.

2

Designed Innovation

…[I]t's really hard to design products by focus groups. A lot of times, people don't know what they want until you show it to them.

Steve Jobs[*]

2.1 INNOVATION USES MORE RIGHT BRAIN THAN LEFT

No one remembers who invented the wheel but whoever it was didn't use any data. In the next case, data analysis is rebalanced with the creative spark used in innovation to help in new product design merchandising and sales.

This rebalancing of data with creative leaps immediately opens up business improvement to both left-brain analytical and right-brain creative people (and people with some of each) so that all can contribute. The competitive power of harnessing everyone's knowledge and skills beyond his or her job was brought to life vividly in the following case, again using statistical design and control but in a very different setting from the opening care management work.

2.2 RETAILING CASE: NEW PRODUCT SALES

The retail case began with a senior management need: increasing sales of a new furniture product line. It began with a serendipitous meeting in the

---

[*] Business Week (May 25, 1998).

hallway about a new product that had been introduced a year previously. Sales were respectable but a continuing economic downturn had reduced sales of all products about a third compared to the previous year. This intensified interest in stimulating sales of this particular new product.

That first meeting also asked whether it would be better to knock the kinks out of the new product rollout before trying to improve it. It made sense though to start before habits were formed (among both customers and employees). It's actually easiest to improve something before it's standardized. Also, that way, any increased sales start sooner and continue to accrue.

The problem was later sized up and scoped out with the merchandising manager over lunch. Sales were measured in each store by units sold monthly, broken down by product line. Previous year data were also captured as sales among stores differed. This allowed different stores to be compared using change since their own previous year sales.

Sales were highly seasonal, peaking in May and November. This would not affect the statistical design already in mind because it would run in a sample of stores all at the same time. Retail sales were through hundreds of stores nationwide, each selling several products from many vendors and manufacturers. Realistically, with such geographic spread, any work to increase sales had to be accepted voluntarily by everyone.

We stopped by an executive's office after the meeting to ask if he could spare five minutes a week to manage the project. A quick check confirmed all departments involved reported to him in some way, therefore the stage was set.

2.3 DISCOVERY

As luck would have it, there was a sales meeting just around the corner with everyone flying in anyway. They assigned a couple of hours on the agenda to explain in 10 minutes how statistical design and control would look and how it would solve their problem, and then get to work. All that work meant was leading through discussions to understand how the sales worked and then starting to list changes that might increase sales.

Cracking the problem had an experienced force behind it throughout, starting with intervention discovery. Some offered to take care of this offline but that would have missed the mark. Good science doesn't start with not being there at the most critical point. Also, good design would be impossible without learning how the thing worked while leading discovery.

Discovery followed, researching changes that might increase sales. An informal research meeting was held with about 20 people who among them knew much of what there was to know about sales, including merchandising, marketing, and advertising.

The session sparked controversy but no consensus was sought. Instead, some of the ideas for change represented conflicting views (e.g., number 3 in the following list) or one small voice (e.g., number 1). The output from the session was a list of potential interventions and these were then fermented over two to three weeks as we probed and adjusted the list with new information and questions. Throwing people and coffee into a room and hoping a good list will emerge is not realistic. It needs some science in it. Here is that raw list. Some items are choices; others (with a question mark) are proposed changes:

1. Which room layout to use in the showroom: focus group or a sales idea proposed in discovery.
2. Photos of the product in real homes in the showroom?
3. Send advertising to past customers?
4. Whether to display custom colors or place glossy foldouts of those colors (like paint charts) in a box.
5. Delivery in 15 days (vs. standard).
6. Store location (geographic region).
7. Store sales volume (larger vs. smaller).
8. Whether to have one or two levels of product (i.e., on the floor and raised by frames).
9. Whether to pepper product lines or keep them clustered.
10. How to price product line #1?
11. How to price product line #2?

The test strategy by now firmly in mind was specific, precise, and deliberate. Over and above local market conditions, the aim was for improvements that worked nationwide in all stores. For this reason, numbers 6 and 7 were dropped. Also, number 5 was found to be impossible to manage, so it too was dropped.

2.4 MEASUREMENT QUALITY

Stores agreed to provide copies of their existing manual sales tracking sheets for a week. These were then tallied and compared to official corporate reports. A couple of anomalies were found, tracing back to how returns were counted, so this was operationally defined at least for the test. One administrative error was found, in a keystroke error that had recorded the previous year's sales far too low. Spot checks were run throughout the test in this way, to verify sales data other than rounding errors. It was important not to add work and just use what was available.

2.5 PREPARING FOR THE TEST

After the discovery session three core people regrouped to assemble the test. Units sold in each store, each month, was the adopted measurement. This was used rather than sales revenue so that price elasticity of demand could be measured.

In the firm's past, it had been hard to convince stores that a particular corporate sales program would work. Even when one did, success had had many fathers. Furthermore, because sales were different in each store, especially because store size and volume varied widely, market tests had been regarded as inconclusive. It was helpful to preview for everyone how all of this would work to instill confidence that the test would solve the problem and also gain broad acceptance in the field.

Sales data for a random sample of 32 stores a year before were therefore sketched on a paper napkin (Figure 2.1). Just as for the care management case, store sales had to be "similar" (really, homogeneous) to support a viable test. The sketch found the data to be skewed; stragglers out to the right had had seemingly unusually high sales. In fact the sketch shows they're not "unusually" high but the whole sample tends to spread out like this to the right. It's certainly not a bell-shaped curve.

There is mathematics that will deal with this and determine whether store sales are homogeneous, but to make this clear to nontechnical people a useful trick is to plot the data on a scale of 1, 10, 100, 1,000 rather than

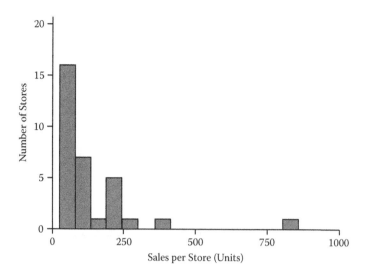

FIGURE 2.1
Unit sales per store for a three-month period one year previously.

100, 200, 300, and so on. That too was sketched, as shown in Figure 2.2.*
This is much more friendly to the eye, without changing anything in how
the underlying mathematics will work in the test to improve sales.

Next, the same data were plotted on another paper napkin in random
order, still on the user-friendly scale marked off in these equal increments
at 1, 10, 100, and 1,000 sales units over three months of the previous year's
sales (Figure 2.3). It's easier to see real outliers if the skew is made easier for
the eye to distinguish. Sure enough, store sales varied widely but only one
store stood out slightly from the rest. Store #29 was the largest store in the
sample and sold the highest number of products. We decided to keep it in
but design the test so we could check that it didn't interfere in the results.

Everyone at the growing meeting (merchandising and design people had
stopped by after awhile to join in) had run market tests to evaluate, say, a
merchandising change in a few dozen stores against a few dozen more as

---

* This is a logarithmic scale which makes a skew dataset with a long tail, closer to a bell shaped
curve, without changing the true story. Otherwise the eye would be misled by stragglers with high
sales when in fact they're to be expected. What's really needed is whether any store is unusually
high and the specially scaled chart does that. The sketch uses base 10 logarithms being easier
for intuitive appreciation than base e but makes no difference to the visual analysis otherwise.
Spreadsheets have this function.

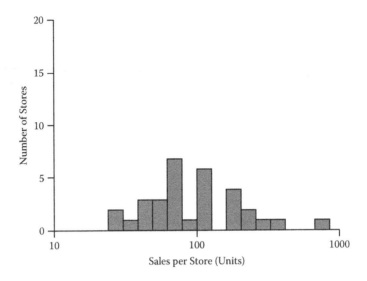

FIGURE 2.2
The same sales data sketched on a 10, 100, and 1,000 scale.

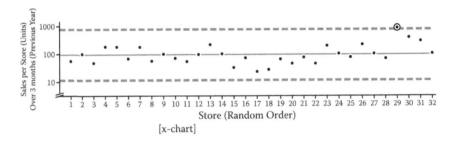

FIGURE 2.3
Store sales are almost homogeneous.

a control group. They also knew it was common sense to pick <u>and</u> split the store sample at random.

A preview of how we would defuse any arguments about the cause(s) of future sales increases was also illustrated, using the napkin sketch shown in Figure 2.3. By ripping it in half vertically, and placing the average for each half as horizontal lines, it was easy to shift the right-hand side up a little (for 16 of the 32 stores) and see the pretend sales increase. The pretend increase was especially clear in the shift in the split average lines.[*]

---

[*] This happens because of the central limit theorem. When the split averages are used, their variation is reduced by $\sqrt{16} = 4$ compared to the variation among the stores (captured in the dotted lines). Essentially this averages out the noise leaving only a slight pre-test bias. The care management case had this going too: in 19 dimensions all overlaid, for the 19 interventions.

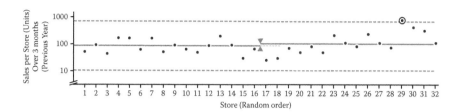

FIGURE 2.4
Pretest bias.

Jiggling the napkin we found we would be able to see fairly slight increases in sales (and larger increases more easily). By reforming the napkin to its original shape (Figure 2.4) we also saw the split average lines were not quite the same but very close. It was easy to see this was just pretest bias in the random selection and we'd have to adjust for that in the final analysis. The whole napkin discussion took about a half hour including plotting the data by hand as someone from sales read the numbers. So far this had used only common sense and arithmetic.

2.6 RETAIL FURNITURE STATISTICAL DESIGN AND ITS MANAGEMENT

Retail experts then prepared (the same day) a standard procedure for each intervention. The format was a page for each with the counterfactual also written down. Meanwhile, this gave a little time for the scientist to catch up with the initial statistical work. This was already shown in Figures 2.1 to 2.4. All this entailed was taking the napkins and a copy of the source sales data and doing the charts again neatly.

The three-month period was chosen to be equal to the intended duration of the testing. Usually this takes a month but the tradition in the brand was for 90-day market tests. In order to be credible, this was maintained. As it turned out though, the results were later clear in a month and that story never changed.

Defusing again the common illusion about randomization (the same one that surrounded the nurses earlier): it might appear that retail stores of differing sales, sales trends, shapes, sizes, geographies, and local economic or competitive conditions (not to mention varying salesperson skills and consumer demographics) are too different to form a "level playing field." In fact, the only criterion is the simple homogeneity test in Figure 2.3.

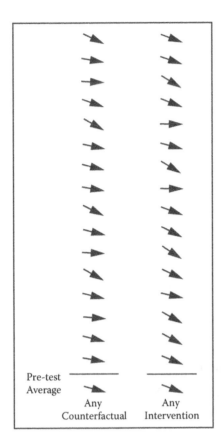

FIGURE 2.5
Sales trends stores were already on randomizes out.

Another easy way to see once again why randomization eliminates all of these things is just to pick a couple: sales volumes and trends that stores were already on, for example. Figures 2.5 and 2.6 show why these will randomize out.

In this case, the illusion came up in a slightly different way and people had concerns. Because area sales representatives (each with a territory containing several stores) varied in their skills, might that not contaminate the test? Although technically not necessary, comparing each intervention to its counterfactual by area sales reps (Table 2.1) made sense to everyone and convinced them that the planned test would be viable. They knew the sales managers and how well each performed and could compare this to the test design now sketched out in rough form (Figure 2.7) and see that no matter which intervention they tried, it seemed to have a reasonable balance of sales skills in its test versus counterfactual. This is as expected

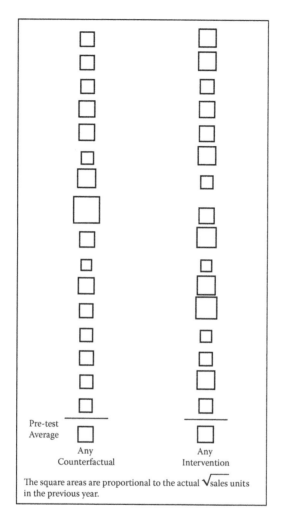

Pre-test Average

Any Counterfactual

Any Intervention

The square areas are proportional to the actual $\sqrt{\text{sales}}$ units in the previous year.

FIGURE 2.6
Store size randomizes out.

under the theory but was still fascinating to watch live. It's a rare glimpse of randomization at work. It's a direct window rather than a more flimsy mathematical model that might be supposed would see if randomization had "worked" (Table 2.1 and Figure 2.7).

A plan for managing adherence to the test procedures was worked out in a short phone call. The firm's local merchandisers, visiting stores regularly, were given a checklist by store. As always, this was designed using intent to treat, to ensure no more management emphasis than could be applied long term at full implementation. Leaping to the statistical design [8] to see

TABLE 2.1

Reassuring Look at Territory Sales Managers'
Skills by Store

Row	Sales Manager	Row	Sales Manager
1	18	1	17
2	15	2	2
3	7	3	3
4	15	4	6
5	6	5	3
6	4	6	12
7	5	7	11
8	16	8	8
9	6	9	14
10	12	10	13
11	12	11	17
12	19	12	10
13	2	13	1
14	16	14	4
15	8	15	9
16	19	16	4

+ test   - not (=counterfactual or "control")

				Interventions						
Row	A	B	C	D	E	F	G	H	Store 1	Store 2
1	-	-	-	-	-	-	-	-	_____	_____
2	+	-	-	-	+	+	+	-	_____	_____
3	-	+	-	-	+	+	-	+	_____	_____
4	+	+	-	-	-	-	+	+	_____	_____
5	-	-	+	-	+	-	+	+	_____	_____
6	+	-	+	-	-	+	-	+	_____	_____
7	-	+	+	-	-	+	+	-	_____	_____
8	+	+	+	-	+	-	-	-	_____	_____
9	-	-	-	+	-	+	+	+	_____	_____
10	+	-	-	+	+	-	-	+	_____	_____
11	-	+	-	+	+	-	+	-	_____	_____
12	+	+	-	+	-	+	-	-	_____	_____
13	-	-	+	+	+	+	-	-	_____	_____
14	+	-	+	+	-	-	+	-	_____	_____
15	-	+	+	+	-	-	-	+	_____	_____
16	+	+	+	+	+	+	+	+	_____	_____

FIGURE 2.7
The furniture sales statistical design.

how it works: for the eight interventions that emerged (labeled A–H), with each of its rows replicated in pairs of stores, the test design with places to record test sales data is shown in Figure 2.7.

Store names are excluded here for confidentiality and were randomized to the test design layout by using the napkin randomization. That also had the advantage that people had seen it done so any mystery surrounding randomization was gone.

Each intervention had by now been written up in language that store people could follow. Thus it was easy to drop the procedures into the Figure 2.7 matrix and print off two each of the 16 test procedures: one for each store. Everyone saw that most stores would test four interventions (the letters above each row's "+"). Had "all +" been cumbersome, we'd have used a different design. Formally, these rows are called *treatment combinations*, but "rows" was simpler for users.

Everyone also saw that each intervention was colloquially "test versus control," most visibly in D because it falls as 8– "control" then 8+ "test." From there it was easy for people to see that C would not contaminate the analysis for D (inasmuch as it would cancel being in perfect balance in clumps of 4– then 4+, etc.). And so on for B and A. People trusted that the same phenomenon occurs for all interventions but that's harder for the eye to see. A couple of people had a nice time re-sorting the matrix in a spreadsheet and making any intervention fall as two blocks of 8–, 8+, and everything rippled through as D, C, B, and A had.

The randomly selected stores were allocated to the test design at random, therefore the chance of something else aligning with an intervention and causing a spurious sales increase in the test would be very small,[*] meaning near zero in practice.

The market test was run with sales data captured each month. The final results were actually clear in the first month and never changed, but were kept under wraps, so as not to bias the remainder of the test. The reason again for the extra couple of months had to do with tradition and what reasonable people would regard as credible. This is another of many aspects where the science can move more quickly than the business.

---

[*] "Very small" is sometimes confused with coincidence (which tends to be more likely). Here, the chance of something else causing a sales increase in the test is about one in 75 million. That comes from how many random splits of 32 stores exist: $^{32}C_{16} = 32!/(16!\ 16!) = (32 \times 31 \times 30 \times \ldots \times 2 \times 1)/(16 \times 15 \times 14 \times \ldots \times 2 \times 1)^2 = 601$ million plus some change. Then divide into 8 interventions gives a 8/600 million = 1/75 million chance of randomization "failing." If this were worth worrying about we'd also buy more lottery tickets!

2.7 EXPLORATORY ANALYSIS AND INFERENCE

The sales test data (and from a year before) are given in Table 2.2. As expected under the recession, sales are down compared to the previous year. Of course this could not affect the test because the recession hurt every store. Notice it is impossible to see that large sales improvements hide behind the array of data.

Analysis using change in sales per store was completed by everyone in an hour or so, using a pencil, just averaging the + versus − for each of the eight interventions. This exercise helped with inference and later findings, more so than everyone simply seeing a slide of results. Analyzing sales change makes common sense as it adjusts for differing sales volume by store and also takes care of that skew data (Figure 2.1).

TABLE 2.2

Test Results: Sales (Units) in 90 Days

	Interventions								Sales Change (Units)	
Row	A	B	C	D	E	F	G	H	Store 1	Store 2
1	−	−	−	−	−	−	−	−	−12	−100
2	+	−	−	−	+	+	+	−	−31	−73
3	−	+	−	−	+	+	−	+	−51	−74
4	+	+	−	−	−	−	+	+	−43	66
5	−	−	+	−	+	−	+	+	−62	−15
6	+	−	+	−	−	+	−	+	30	−38
7	−	+	+	−	−	+	+	−	−57	−62
8	+	+	+	−	+	−	−	−	178	22
9	−	−	−	+	−	+	+	+	−3	20
10	+	−	−	+	+	−	−	+	−16	36
11	−	+	−	+	+	−	+	−	−8	−4
12	+	+	−	+	−	+	−	−	82	−24
13	−	−	+	+	+	+	−	−	−28	−150
14	+	−	+	+	−	−	+	−	−25	16
15	−	+	+	+	−	−	−	+	−275	−222
16	+	+	+	+	+	+	+	+	−73	29

Exercise 8

Analyze Table 2.2's data to find the effect of each intervention. Start with D (which is easiest) and average all the data in the last eight rows (D+); then subtract the average for the first eight rows (D–). This gives the effect of D on sales change. Then repeat for all the interventions following the + and – signs in each column. It's easier to start by averaging the pairs in each row and just use those. By comparing to the condensed recap of the interventions in Table 2.3, think through what each result means. The answers appear in Table 2.4.

Table 2.4 gives the results, augmented with the average sales change for each row's pair of stores, and their corresponding range (i.e., by how much their sales changes differ).

To illustrate how this design also looks at pairs of interventions that do well together, Figure 2.8 reorganizes the data into a 2×2 on the pair of interventions: A and G.

The choice of AG* arose immediately once the exploratory results were complete. Interventions with large effects tend to interact with other interventions. This phenomenon is known as heredity of effects [6]. Because A has the largest effect it's under suspicion for interacting with other interventions. All such 2×2s were tried and this one was large. The plot is largely self-explanatory, suggesting A + G– gives the highest sales.

This one's now ready for us to translate back to the real retail world, understand what the results mean, and then develop findings to implement.

---

* Statisticians will notice the design used here was a 2^{8-4} Resolution IV, one-sixteenth fraction factorial with a simple aliasing scheme. AG is therefore aliased with CD, BH, and EF. Although heredity makes AG most likely (and the physics of it, that marketing people realized, strengthens the notion of AG being real) it has to be verified in refining testing or by planned implementation, building in a 2×2 in A and G. Because there was no time or field interest in this, the final implementation plan was carefully chosen so that even if one or more of the aliased interactions were real (as well or instead) the implementation would mostly capture the benefit(s). This "can't lose" strategy has been found useful where further testing is not realistic from the business angle.

TABLE 2.3

Condensed Recap of Interventions

	Interventions	Counterfactual (–)	Test (+)
A:	Layout	Focus group's	Sales'
B:	Product photos in real houses	No	Yes
C:	Advertise past customers?	No	Yes
D:	Custom colors	Samples	Glossy color chart
E:	Print price	–10%	+10%
F:	Plain price	–10%	+10%
G:	Pepper prints among plains?	No	Yes
H:	Levels displayed	1	2

TABLE 2.4

Exploratory Findings

	Interventions								Sales Change (Units)			
Row	A	B	C	D	E	F	G	H	Store 1	Store 2	Average	Range
1	–	–	–	–	–	–	–	–	–12	–100	–56	88
2	+	–	–	–	+	+	+	–	–31	–73	–52	42
3	–	+	–	–	+	+	–	+	–51	–74	–62.5	23
4	+	+	–	–	–	–	+	+	–43	66	11.5	109
5	–	–	+	–	+	–	+	+	–62	–15	–38.5	47
6	+	–	+	–	–	+	–	+	30	–38	–4	68
7	–	+	+	–	–	+	+	–	–57	–62	–59.5	5
8	+	+	+	–	+	–	–	–	178	22	100	156
9	–	–	–	+	–	+	+	+	–3	20	8.5	23
10	+	–	–	+	+	–	–	+	–16	36	10	52
11	–	+	–	+	+	–	+	–	–8	–4	–6	4
12	+	+	–	+	–	+	–	–	82	–24	29	106
13	–	–	+	+	+	+	–	–	–28	–150	–89	122
14	+	–	+	+	–	–	+	–	–25	16	–4.5	41
15	–	+	+	+	–	–	–	+	–275	–222	–248.5	53
16	+	+	+	+	+	+	+	+	–73	29	–22	102
Effect	77.4	–4.1	–31.1	–20.2	20.4	–2.4	19.8	–25.9				

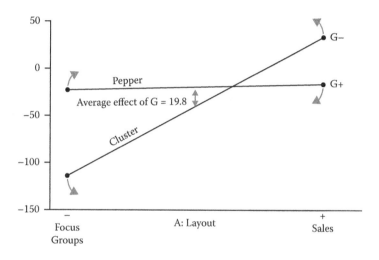

FIGURE 2.8
2×2 and interaction plot for A and G.

2.8 WHAT MIGHT THE RESULTS MEAN?

The first clue that something special had happened was that almost all of the hundreds of stores not in the test saw sales drop an average of 37% from the previous year due to the recession, whereas among the test stores, about a quarter saw increased sales, and among the 32 test stores overall, the drop due to the recession was less, at an average of 22.1%. This carried interest as the 32 stores had been selected at random.

Improvement like this, during the experiment, has been seen more often than not in tests involving people. It's the standardizing effect of controlling so many variables. This is another reason for a light-touch, intent-to-treat approach to test management: if it's too heavy it could cause a Hawthorne or "spotlight" effect just from senior management attention to a few stores. There may still be some of this but it would not be expected to sustain for three months. Either way the interventions are measured pure, because they enjoy the same management attention. There was more to come at the implementation of the test's findings, but this was a start.

Exploratory findings were reviewed with merchandising staff who were perturbed because the layout tested out opposite to what had been expected. The idea to try a salesperson's idea had originally been controversial as market research had already designed the best layout using focus groups. It had been expected the test would re-prove this. Yet the focus group design had been outperformed. After reassuring staff that the result was real, there was a call with the executive who'd agreed to sponsor the effort. The important thing was to find out through the eyes and ears on the ground what had happened. These belonged to the salesforce: both the territory managers and the salespeople in the stores. The request to the executive was to support getting the word out to the field and asking them what they knew.

Innovation is rarely flocked behind at first, so the controversy heightened, with some saying the whole test had failed because this layout result could not be right. This is often a volatile mix and one that can't be completely controlled. The place to start and stay is at the top, but at the same time avoiding a top-down edict, inasmuch as detective work such as this goes wherever it goes and no one knows the correct explanation yet. Without senior management support, the project would have perhaps been killed at this juncture.

Within a couple of weeks, word started to trickle back from the field. The controversy and credibility problems started to crack and, starting like a whisper, momentum built for the innovation. It built on the grapevine at first. The explanation had to do with how people shopped and that the amateurish design made it easier for consumers to find what they wanted. How it differed from the official layout was in its more informal appearance; it was also more compact. In addition, it was assembled by the store salespeople whereas the research design had to be assembled by visiting merchandising specialists.

The focus groups that had liked the official design had been previous and potential customers. Therefore the finding revealed that customers had erred in assessing how they, themselves, would most shop. In this kind of problem, customers have typically been found no more correct than the rest of us about their wants and tastes; they just knew what they liked when they saw it and chose to buy it.

The remaining results were less surprising but with considerable further business advantage: placing product photographs from customer homes (B) was immaterial, being almost perfectly flat on sales. Advertising to past customers (C) by sending a circular to their homes had been controversial. Some felt it would create the wrong impression inasmuch as this was a quality product and not the least expensive on the market. Others felt it important as the product could easily be added onto with additional units in the collection. At any rate it did not help sales and if anything it leaned to supporting the view that promoting in this way was not appropriate, being the largest hurtful intervention at face value during this exploratory step.

The glossy color chart (D) may have hurt a little so the tentative decision was to use the samples to show color and print options for the coverings. The two price elasticity interventions (E and F) received much attention. Neither had hurt demand (with F admittedly slightly negative but the smallest in the list and very close to zero). If anything, E had increased demand slightly. The final decision awaited calculations of statistical significance, with a tentative position that the higher price made sense. This made other sense too, because the product was fairly high quality and it was known customers were not primarily shopping on price.

At face value, the floor layout peppering both products together (G+) looked promising, with a slight sales increase associated, but this was countered by the interaction (Figure 2.8) suggesting there was more to G than first met the eye. Again, the meeting agreed to wait for final significance calculations but asked what a potential AG interaction would mean and if so how large it was. This was easy to do in real time during the meeting. Armed with Figure 2.8 as a penciled sketch, the answers were explained as follows.

An interaction just means that the effect of G depends on A. Figure 2.8 shows this because G+ hurts with the A+ layout but helps with the A− layout. So the effect of G depends on the choice of A.

The twist exerted by this significant interaction is illustrated in Figure 2.8 with the curved arrows. A conversation to help understand what this means physically might be:

Person 1: What are we getting out of G?
Person 2: It depends.

Person 1: On what?

Person 2: On A.

Person 1: How so?

Person 2: Well you can see in Figure 2.8 that if you choose A+, then G+ hurts relative to G–, but if you choose A– then G+ really helps.

Person 1: So what should we do?

Person 2: A+G–.

As for the size of the interaction, it's visually clear in Figure 2.8 that AG is larger than G. This AG interaction therefore clearly becomes the dominant consideration over G alone. Method 1 quantifies AG. The meaning of the AG effect being negative at –70.3 is a little more complex to interpret

METHOD 1: A SIMPLE WAY TO CALCULATE INTERACTION SIZE

Just subtract diagonal averages on the 2×2:

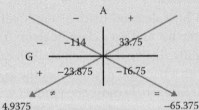

averages: 4.9375   –65.375

The diagonals are labeled, by whether A and G's ± signs are the same (=) or not (≠).

The convention is ↘= – ↙≠

= –65.375 –4.9375

≅ –70.3

This can be done instantly by eye in meetings as questions come up.

It also reflects how the mathematics works (Chapter 5).

Checking Figure 2.9 gives zero.

Hypothetically plotted points (which preserve the numerical story) were:

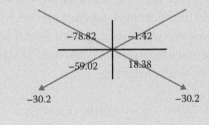

–30.2 – (–30.2) = 0   averages: –30.2   –30.2

∴

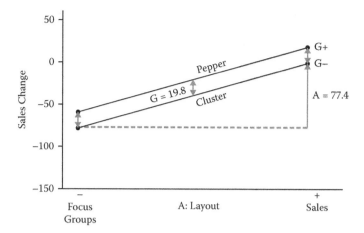

FIGURE 2.9
How zero interaction would look: one hypothetical example.

than a single intervention, so a plot is always advisable. No one can interpret what the –70.3 means without a plot.

Having only one level of product displayed (H) was a quick unanimous decision that will not change once significance is quantified. One level of product was also easier to manage.

The informal discussions continued for a few days with most interest centered on the surprising layout finding. The executive had asked for an update in a couple of weeks and there was also curiosity about how this could have happened. The credibility of the design had strengthened at the meeting where everything else made sense or brought clear sales advantage.

Now that energy was high in the field (and with the test highly credible again), the discussion quickly figured out why this was so … with hindsight. Had someone tried to explain such a phenomenon with foresight they'd have been thought ridiculous.

Suddenly, then, it was rather obvious. The explanation came back from the stores. It had to do with how the eye visually absorbed information. The amateur design presented several little arrangements in modules like small rooms in someone's home, only one of which could be looked at at a time. The clustering by product line was important so that shoppers could find the style and configuration they were looking for, then choose which one they wanted. If the product lines were peppered, then they had to look at one layout and keep that memory alive as they walked across the store to comparable similar layouts. No one thought of this in advance but it was obvious with hindsight once the test results brought it up.

Conversely, the peppered arrangement worked better with the focus group design because everything could be seen at once, the way it worked: more like someone's home without the walls. Therefore the amateur display worked especially well with the clustered product lines, but clustering product in the focus group layout damaged sales. This latter might well have been the decision, without testing.

2.9 STATISTICAL SIGNIFICANCE

Figure 2.10 gives the normal probability plot as used in the care management case. It makes clear that A and the AG interaction stand out, with a

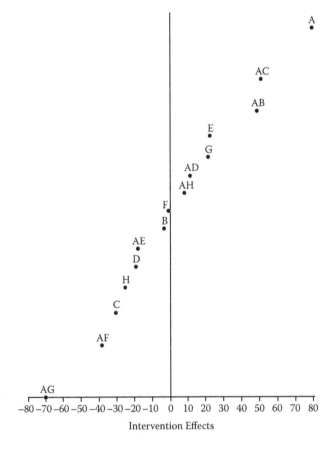

FIGURE 2.10
Plot of intervention and pair interaction effects.

few more interactions unclear, whereas nothing else appears off the rough straight line in the center of the plot.

All seven interaction effects with A that the design can estimate are also shown. These can be verified by preparing all the 2×2s inheritable from A, as for AG.

Exercise 9

The straight line is split near its middle. Why? What valuable use could that be put to, in all statistical designs?

METHOD 2: IMPORTANT SHORTCUT FOR CALCULATING STANDARD DEVIATION

Most shortcuts give a quick approximation by cutting corners.

The following shortcut gives more "exact" answers by cutting uncontrolled causes of variation: here the interventions themselves.

It's obvious that Table 2.4's data cannot estimate standard deviation ($\hat{\sigma}$) by plugging all the data into a calculation, because going down the page, interventions are changing. Across the page though, interventions are held constant. Therefore the ranges (being the larger minus the smaller sales) give a direct estimate of inherent variation. One range (R) is flimsy but 16 averaged isn't. Using something called the distribution of the range, it turns out that:

$$\text{Upper limit on } R = 3.27 \, \overline{R}$$

$$\hat{\sigma} = \overline{R}/1.128$$

The limit checks if any data are aberrant giving any R too large (which would then be correctly excluded from $\hat{\sigma}$).

The formulae only work for two stores. That covers all cases in this book.

* The constants 1.128 and 3.27 are from d_2 and D_4 tables [9], which give the same constants for all sample sizes rather than just the two (stores) used here.

For this design, significance also can be calculated, giving a more objective criterion than the eyeball assessment. This is because the design was replicated by putting two stores in each row.

METHOD 3: SIMPLE SIGNIFICANCE CALCULATION

For two test units per row, controlling to a 1/20 false alarm:

R is the difference (ignoring the sign) between test units measured for each row.

$$^*\text{Noise} = \pm \, t\, \hat{\sigma}\sqrt{4/N}$$

± (pronounced "plus or minus") means we're interested in negative effects (which would hurt sales) as well as positive ones (helping sales).

t is the t-statistic, with degrees of freedom = number of rows replicated.

$$\hat{\sigma} = \bar{R}/1.128$$

N is the total number of test units (= 16 rows × 2 stores = 32).

Intervention (or interaction) effects larger than the noise are significant.

Any ranges exceeding 3.27 $\bar{R}$ (meaning homogeneity has been lost) should first be excluded from $\bar{R}$, recalculating $\bar{R}$ successively until remaining ranges are below 3.27 $\bar{R}$.

Optional Technical Notes: The common $\hat{\sigma}$ calculation can't be used because conditions change by row. The range method is not much less efficient and brings strong advantage in rechecking homogeneity and providing correct adjustment as above.

* Statisticians can reproduce this formula in a couple of lines by using the central limit theorem when the + and – rows are averaged per intervention, and adding variances when the – is subtracted from the +, landing on the applicable t-statistic as above.

CALCULATION METHOD 2 AND 3

From Table 2.4, $\overline{R}$ = 65.0625 therefore 3.27 $\overline{R}$ = 212.8.

This gives a valuable check on whether any of the stores had unusually extreme sales that could throw the analysis. Glancing back at Table 2.4 finds no range above 212.8, as hoped for. This check is always done where replication is used.

$$\hat{\sigma} = \overline{R}/1.128$$

$$= 65.0625/1.128 = 57.7$$

$$\text{noise} = \pm t\hat{\sigma}\sqrt{(4/N)}$$

t is the *t*-statistic which is looked up in statistical tables [8].

N is the total number of test units: here 32.

Calculating the noise then, *t* with 16 degrees of freedom is 2.12. The degrees of freedom is just how many independent pieces of information went into the calculation for $\hat{\sigma}$. It appears this would be 32, with 32 stores, but in fact they're arranged in 16 pairs and the 16 ranges were used to estimate $\hat{\sigma}$. An argument that there are still $16 \times 2 = 32$ bits of information falls apart because the average of each pair was already calculated (Table 2.4) and once that's known, knowledge of either store's sales reveals the other's. So the degrees of freedom are really $16 \times (2 - 1) = 16$.

Thus, noise $= \pm 2.12(57.7)\sqrt{(4/32)} = \pm 43.2$.

Anything smaller than this calculated ±43.2 can be thought of as noise (not significant). Anything larger is a signal (significant), meaning it's taken as real and likely to affect sales.

Significance is analogous to listening to someone talk about statistics in a noisy restaurant. The ear filters out the useful information (signal) above the noise of everyone else talking in the restaurant. Of course the other conversations are not really noise; presumably they also contain important information. The point is there are too many relatively indecipherable conversations going on to extract any useful information. So we literally call that noise. We should be interested in some of the noise, rather than just significance per se. This is because nature does not work like an on–off switch determined by significance when we're looking to improve a business process. In the restaurant chatter analogy, this means we might pick up something useful by listening in on the next table's conversation. Still, significance is an important part of the scientific method in giving scientific feedback we can rely upon to make real changes that bring real improvement.

This confirms A and AG are highly significant. Two other interactions are just significant and these hold proportionately less interest.

This means A helping at 77.4 was expected by chance less than 0.1% of the time. It would tend to occur once in every 1,000 tests on A. This tends to thwart the inherent concern that when testing about 20 things (here 15 including interactions) perhaps one in 20 is expected significant by chance (with one in 40 helping). That might throw new light on AB and AC, but not much on A or AG.

Putting significance in perspective, it's an important analytical start to the post-test scientific work, which has to be completed. The staff and field discussions in the furniture sales case were useful in that work.

2.10 IRONING OUT SOME POSSIBLE WRINKLES

In the homogeneity check, store #29 showed as a signal and traced to the largest sales-volume store. It might appear correct to exclude it from the calculations (had that rule been set in advance). The reason for defining the exclusion before the test results are in is so that rationalization does not become a great game. In this case we decided early it was so slight we would keep it in. Everything changes over time therefore the most current information as to whether #29 really was aberrant comes during the test.

The upper bound on the range in sales was used earlier. With store #29 paired with another store, their range was 53, well within the calculated upper limit of 212.8. Therefore it was safe and correct to include store #29.*

The dry run was also important. Although stores were found to have homogeneous sales, the pretest bias will always be there for every inter-

---

* If #29 were excluded, regression analysis would find A = 74.1 and AG = –67, which is further reassuring. There is no simple calculation to impute #29 (if missing or excluded), as Figure 1.2 makes clear, so regression is used.

TRANSLATION 3

Regression is a statistical method that finds which variables are related to an outcome measurement. Such relations may or may not be causal. When used to analyze interventions in statistical design, causality is established directly.

vention. It's possible (though unlikely) that A and AG are just picking up haphazard patterns in how the store sales differ. The dry run found:

$$\text{Pretest bias in A} = -57.8$$

$$\text{Pretest bias in AG} = 103.3$$

This reinforces both findings as real, and if anything, understated. In Chapter 5 the data are provided to reproduce all calculations, and with a little more analytical insight.

2.11 PREDICTING AND DELIVERING THE IMPROVEMENT

The implemented interventions were: A, E, and F but not G, which translated to the amateur sales layout, the higher price for both products, and sales' idea of the product lines kept apart in the showroom. It's next to impossible to predict sales at full implementation of the test's findings, given the economic downturn that at some point would reverse.

Arithmetically, it's clear that the increases ought to be large enough to reverse the effect of the recession. In most cases (such as in the care management case) the prediction can be verified. In this case, the only way to validate is by a random holdout sample of stores. Clearly there are problems asking a few dozen stores to delay increasing their sales (and no one knows if that in itself could have some other negative effect on morale and sales). The way this one happened was a decision to let the stores choose. About 80% wanted the test's findings for their stores.

A year later, the implementing stores averaged 9.8% higher than the holdout group. Notice, however, that the 20% of stores declining were self-selected and that will usually carry a bias. This was easy to resolve by finding no such advantage a year prior.

METHOD 4: PREDICTING THE IMPROVEMENT

Improvement is easily predicted from the test's average plus half of each significant effect:

$$
\begin{array}{lll}
\text{Test average} & : & -30.21875 \\
\text{A+} & : & +(77.4)/2 \\
\text{A+G-} & : & -(-70.3)/2 \\
\hline
\text{Prediction} & & 43.6
\end{array}
$$

The halves apply since half of the test's rows tested each intervention (and each interaction), already reflected in the test's average, so these cannot be double counted.

The sign before each entry represents the implementation plan. A is easy to see as it must be + to gain the sales increase. A+G– was decided from the interaction plot so the AG interaction is set at + × – = –

Calculating all four options in A and G clarifies the prediction method and is also reassuring in reproducing the actual test results in Figure 2.8's 2 × 2, quite closely:

	Test Average	A	AG	Prediction	Actual
A–G– :	–30.21875	–(77.4)/2	+(–70.3)/2 =	–104.1	–114.0
A+G– :	–30.21875	+(77.4)/2	–(–70.3)/2 =	43.6	33.8 implementation
A–G+ :	–30.21875	–(77.4)/2	–(–70.3)/2 =	–33.8	–23.9
A+G+ :	–30.21875	+(77.4)/2	+(–70.3)/2 =	–26.7	–16.8

Notice A+G– is predicted to overcome the recession (reflected in the test average).

It could be argued two other interactions be included but these are only just significant and taking into account the normal plot (Figure 2.10) with split parallel lines (explained in Chapter 5).

It's often thought the proof of the pudding is in the implementation but of course it was in the test that ran under rigidly controlled conditions while allowing an element of laissez-faire to ensure its findings were real. It would make no sense to take a pudding proof from a completely uncontrolled happenstance economy. It's important to deliver, but the best way to do that is to follow through on the scientific method and ensure implementation. It was helpful that this sales case was one of several the firm had running in different operations so they could see the portfolio of results at implementation, most of which could be verified. This gave some confidence that the sales case also would deliver. The 9.8% (not found a year previous) confirmed everything was as it should be in implementation. Market share and customer satisfaction measurements also gave steady support to the implementation success.

An aspect that was easier to verify, however, was the additional margin from increasing the price 10%. No one had any doubt about that as the financial statements showed. The year-over-year contribution of the product line was calculated to be over 20% higher.

2.12 RETAILING DESIGNED INNOVATION CASE: CONCLUSION

Not normally thought of as a quality or scientific problem, sales (net of returns) perhaps represents one of the most direct measures of customer satisfaction. The hundreds of stores also presented an unforgiving environment for the scientific method, being geographically spread over thousands of miles.

The case took everyone to a place no one would otherwise have found, including apparently the customers in the earlier focus groups. This facility of statistical design to find solutions no one can visualize (at the outset) has now been seen in the two cases several times. It turns out to be more the norm than the exception in large-scale statistical design.

Ordinarily a few dozen stores had been used in market tests so the smaller sample, ample given the statistical design, saved about two thirds of the usual market testing costs.

The innovation could not be copied by the competition as the solution was unique to that retail brand. Even studying the solution in store visits would not have revealed how it was done.

3

Statistical Control

[T]he object ... is to set up economic ways ... of satisfying human wants ... requiring a minimum amount of human effort.

Walter Shewhart*

3.1 USING STATISTICAL CONTROL

Statistical control used in the cases thus far illustrates its main uses as follows:

- To establish homogeneity among test units (e.g., nurses, retail stores) to support a viable statistical design
- To provide a real-time feedback control system to manage both sudden and sustained improvement and track it (clearly, at a glance)

Both uses apply to both the process (e.g., admit rate, retail sales) as well as its measurement error from the outset.

Statistical control is easy to use and is self-teaching after making a start on real process improvement, with a single expert able to guide dozens of users comfortably. Modern software often has to be tricked into proper use of statistical control.

---

* Shewhart, W.A. (1931, 1980). *Economic Control of Quality of Manufactured Product.* Reprinted 1980. Milwaukee, WI: American Society for Quality Control.

3.2 ECONOMIC ADVANTAGE

The brilliant development [2] of economic control extends over nearly 300 pages before its seemingly simple criterion appears for the first time:

$$\overline{\Theta} \pm 3\sigma_\Theta$$

which translates to "the average give or take three standard deviations" of whatever is being plotted. This gives the dotted lines that appear on each chart thus far, called a *control chart*. The dotted lines are called *control limits*. That the development took as many pages indicates its depth, which only a few technical leaders need to know in full.

The loss of the original term "economic control" is unfortunate. The whole point is economic advantage at all levels, starting at the top one. In the opening pages of the development appears a rather remarkable preview of what the theory accomplishes. Ten similar businesses achieved statistical control to varying degrees, and a weekly chart of the same type as used in the care management and retail cases is shown for each. What happens across the 10 businesses is striking. Figure 3.1 uses those original data to clarify the economic benefit of statistical control. Businesses most often meeting the criterion make the most profit. The principle is general and applies to any business's financial results (not just percent defective as the main cost driver in the preview case) as well as processes upstream and at lower levels.

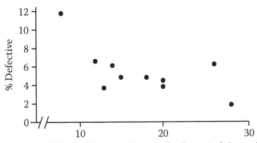

FIGURE 3.1
The intent and outcome is economic.

3.3 DERIVATION

Statistical control [2,10] has been controversial, without reason other than unfamiliarity with its original development. The mathematics is so easy that it belies the correct ingenuity of the theory. It begins with a surprising premise that, like randomization, has no mathematical notation or simple explanation: there is no mathematical statistics explaining the observation (Figure 3.1) that achieving a state of statistical control improves performance, measured by the quality of process output.

Early in the development then appears the simple criterion that proof of the theory is that "it works." This is the same unforgiving criterion adopted here to prove a statistical design (which also has much theory behind it) was done correctly: something improves suddenly then sustains.

At the center of statistical control theory is Tchebychev's inequality. This can be derived in a few lines and, for statistical control, proves that for any data (bell-shaped or not), more than 8/9 of plotted points will fall inside control limits. However, this requires that all of the data be used to calculate the standard deviation.

Statistical control theory tries to plot data in such a way that they roughly follow a bell-shaped curve. This is not necessary, given Tchebychev's inequality, but brings the advantage that the fraction of points outside limits can be predicted more closely. Specifically, for a truly bell-shaped curve, 0.27% of the points will be outside limits: about 3 in 1,000.

The crux of statistical control though is that short-term variation is used as a yardstick for longer-term variation. That is, control limits are calculated using hourly, daily, weekly, or monthly data rather than all of them. Thus in the care management example, weekly data were used in the time-based chart. The same principles apply in the homogeneity checks (that use the same theory of statistical control).

The calculation for standard deviation used in software, spreadsheets, and in basic texts or courses will not work for statistical control. The reason is it mixes natural variation with "spikes" caused by particular changes in operations. Therefore it tends to overestimate standard deviation and set the control limits too wide. The longer the time frame of the data used, the worse will be this error. Thus it makes sense to use shorter-term estimates and avoid too many spikes or undulations.

Although the standard formula can be used specially in statistical control, the range-based methods for variables data (and equivalent methods for proportions, counts, and rates data) are safer and easier. In the care management case then, weekly data were used to calculate the control limits, but not just by slapping them all into the calculations. Instead what's needed is the variational equivalent of drive distance without the detours after the road is fixed.

It makes sense that short-term variation (e.g., by the day) is less susceptible to spikes. Thus by estimating standard deviation from a few days then removing any spikes in the way used already with the upper limit on the range, it will tend to the true natural variation of the process. It is this that makes the range-based methods so useful, and overcomes their slight inefficiency compared to the more widely known method. Therefore, far from being a primitive inferior method, this advantage trumps in the real world.

The control limits, estimated in this way, become predictive of what can be expected in the near future and therefore as a yardstick for when new spikes appear. In fact, there are more than spikes (such as undulations, trends, and yo-yoing) that are added to the mix shortly.

Statisticians describe the situation of statistical control as using variation within (say) weeks as a yardstick for variation among them. Using the variation among weeks to estimate standard deviation will not be correct in the real world, or even approximately so, in most real business cases.

In no sense are spikes being nefariously censored so as to present a better picture of things. They are being excluded from the calculations to see how the natural variation looks. The spikes (and other assorted unnatural patterns) remain in full view, on the chart plotted. They just do not belong in the calculations of process average and control limits. The process therefore divulges its own spikes among its natural variations. There are no exceptions to this in any industry, process, or problem type.

The Tchebychev inequality provides something of a safety net where the subtending data plotted are not quite bell-shaped, or even close. The control limits will still do their job of guiding economic action to bring economic advantage and optimality given the current process design.

Returning to Figure 3.1, the original charts covered 36 weeks of running the businesses. Therefore the fractions of data outside limits are readily seen to fall between 8/36 and 28/36. This apparent violation of Tchebychev is because the short-term variation is correctly used to calculate control limits. This gives a reasonable feel for what can be expected in any process.

3.4 PRACTICAL USE OF STATISTICAL CONTROL

The strategy, directly from statistical control theory, is to standardize (by using statistical control) then improve (e.g., using statistical design). This was done in the care management case but not much discussed yet.

The first three weeks, after starting work on the care management improvement, were used to standardize. Most of this work was in gaining statistical control on measurement error. Actions taken were economic (by working only on measurement error falling outside its limits) and stopped short of becoming uneconomic (by eschewing "perfection"). It is important to note that the measurement error itself gave the information as to how it could be improved and a nice illustration of how problems can be exploited to solve themselves.

In a similar case, statistical control found one nurse's admit rates much higher and considerably above the upper limit. This guided discussion and data to find out why and use the information. The reason was the particular nurse's skill with a specific health condition. Noticing this, managers had assigned such patients accordingly. The nurse explained that when such patients were hospitalized, perhaps for surgery, they had to be taken off their medications, because of a clash with anesthetic. After discharge, restabilizing back onto the original medications could be difficult, leading to readmission. For the statistical design about to start, it was easy to exclude that nurse and use a "spare" nurse. Perhaps more importantly, it triggered the idea for a new intervention to test and reduce these readmits.

On the third Monday after work started on Chapter 1's case, the statistical design began, having first ensured statistical control on the admit rates over time and among nurses (homogeneity) as well as on measurement error (also both temporally and by nurse).

As an aside, interestingly enough, some processes cannot be standardized in the sense of all points plotted over time then falling within limits. An example would be retail sales traversing an economic shock, where Chapter 2 gave a rare case in which statistical control could not easily be employed over time, because sales had shot through the lower limit by a long way due to the recession. In that case, statistical control was largely ignored temporally (for awhile) but exploited among retail stores. (In fact statistical control has been used in retail in its more advanced adaptations to link to retail and competitive indices but these are beyond the scope of this practical book.)

This opens up a vital element of improving processes: when looked for, statistical control will always be found somewhere. Once that inherent statistical control is found (or managed), then improvement is assured just by completing the improvement work.

Another example where statistical control over time is unrealistic is telecommunications and other utilities where lines are exposed to weather and thus outages, repair times, and the like increase in bad weather. Because weather happens everywhere on the planet, it is sometimes only possible to find statistical control among central offices, technician crews, and other spaces that move through time together, in statistical lockstep.

The care management case was brought into statistical control over time, whereas some processes do not and therefore jump straight to statistical design. The measurement error step can never be missed, however. That has to be brought into statistical control at acceptable quality.

The most intense use of statistical control came just after the statistical design was implemented and for a few weeks after that. Figure 1.12 wasn't given much discussion in Chapter 1 other than to confirm the improvement and that the admit rate by then varied less over time (which also comes with the theory). The fuller story was that as implementation occurred, the work started to overcome initial nurse trepidation as already narrated. Statistical control revealed to the scientist what was occurring. Statistical controls were kept on several splits of the main cohorts shown in Figure 1.12. These warned the improvement was not emerging as it should have been in the first few weeks or so.

The critical phase occurred from about Week 14 and for a couple of months. Figure 1.12 shows the admit rate almost touching the lower limit once and, coupled with Figure 1.13 not approaching the lower limit; this is valuable information indeed. So much so that it deserves its own section after a digression first.

3.5 DIGRESSION INTO CAUSALITY

A way to see statistical control is Figure 3.2. Data points plotted outside limits are from causes economic to pursue today. Causes will be easily found and either standardized in (if good), prevented (if bad), or ignored (e.g., if due to an ice storm shutting down operations). In practice they will

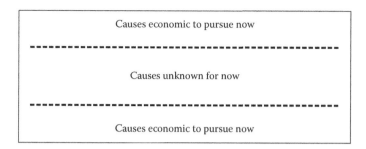

FIGURE 3.2
Statistical control and causality.

usually be found almost immediately, with a little detective work helped by the people in the trenches, who are the eyes and ears on the ground.

Conversely, data points plotted within limits will be from causes too small, complex, or many to understand today (and some never). In other words they will be uneconomic to pursue at face value (meaning, as if they contained usable information). Statistical design later finds some of those causes by ignoring them and leaping to solutions directly. Other sophisticated analyses can also be used to untangle what's inside limits. However, a good strategy is to start in the easy places: with points outside limits.

Accordingly, the variation within limits is not random (although it may follow random patterns).

Once a process is maintained within limits by management actions and an improvement effort is underway, the limits give a reliable test for when improvement has occurred. There are actually a few tests but the limits are the most stringent.

In the care management case, the admit rate over time, pre-improvement, maintained reasonably within limits, and approximated a bell-shaped curve quite well. (It's actually something called Poisson but this is well known to approximate the bell shape quite closely where admits are rare events and a lot of patients are pooled.) It's also well known (or easy to look up in statistical tables) that 0.27% of data from a bell-shaped curve will fall outside limits (set at the average give or take three standard deviations). Therefore, in an improvement effort, the process would be expected to fall below the lower limit naturally (inasmuch as that's the only limit being aimed at) half of 0.27% = 0.135% of the time. For data plotted weekly, this equates to 0.135% of the weeks or 0.00135 expressed as a proportion. This means typically about once every $1/0.00135 = 741$ weeks the process would naturally break the lower limit. Re-expressed in years, $741/52 = 14.2$ years.

So a practical way to look at this is that once brought into statistical control, the process would naturally break the lower limit about once every 14 years. To that extent it's been helpful to nontechnical people to think of this as a "14-year flood." This gives a better sense for how unlikely it is that a single point will cross the lower limit given statistical control has been reasonably established by standardization.

Of course, at the outset, Figure 3.1 indicates the limits will be crossed quite often. In the care management case they were, especially due to measurement error, which was of such poor quality the perception was one of its being unusable. After gaining reasonable statistical control, crossing limits naturally will be rare.

Single data points outside limits are perfectly adequate to gain the economic advantage of statistical control, whether during the early detective work to exploit them and standardize most efficiently for now, or more so when looking for the first clue as to emerging fledgling improvement. The reason for this perhaps surprising reliance on single data points is that we're not. We're using the exclusion of the single point from the many that preceded it. Therefore the exclusion rather than the "singleness" is what's usable. This is similar to how radar finds tiny signals in large amounts of noise to track planes safely. No one cares that each data blip on the screen is from extremely rapid sampling, comparing tiny signals to large amounts of noise. It did, however, take awhile at first before people trusted radar and agreed to fly on planes that used it, without eyeball confirmation.

This takes a little usage to get accustomed to, but after a short while, users find most points plotted outside limits yield easily to their detective work and the improvement keeps creeping up, visibly, behind it. The care management case is about ready to be returned to, keeping firmly in mind that it will make use of single data points for awhile, until as many as 10 data points in a row confirm the single ones had been on the right track. Single data points, when using statistical control, stick out like a sore thumb from a bunch of fingers.

3.6 CONCLUDING SCIENTIFIC WORK IN THE CARE MANAGEMENT CASE

Without any further ado, the crux of the care management case can be described. There seems to be a pivot in each improvement effort where the case is made or broken. Of course the statistical design drove the improvement eventually shown in Figure 1.12 but those are generally easy to run. A little trickier was making sure everyone stayed implementing.

Following the meeting with the nurses to increase adherence to the statistical design's findings from about half (with one intervention at 20%) to most of the time, the improvement took hold and became visible by Week 15. Not quite at the statistical control limit but very close (nature doesn't work as an on–off switch). It's almost a "14-year flood." If that were all we'd had it would have signaled improvement was on its way. It can be put in more perspective by noticing that the full improvement predicted from the statistical design was at about a quarter down from the prior average and Week 15 is below that. This also gives the insight that the fully improved process will not routinely break the lower limit so the other tests for statistical control will shortly be needed to confirm.

As bad luck would have it, preparations had been underway to increase the number of members in care from about 3,000 to over 5,000. By Week 15's emerging good news they were starting to arrive in increasingly large numbers each week ready for assignment to care. The existing nurses (numbering a little over the 20 used in the statistical design) were being complemented with new hires. As that occurred, experienced nurses helped train the new nurses and the care management program was curtailed for a few weeks to free up the training time as experienced nurses mentored new nurses. At the peak of this increase in capacity, the care management program was essentially out of action.

Not surprisingly, the admit rate crept up and broke the upper limit. Even less surprisingly this caused some angst. Some wondered if the statistical design had by now broken. At the time, the science of it was not accepted by everyone, with perhaps half seeing it. Without statistical control and Week 15, it's likely no one would have seen it. This had to be translated frequently by the scientist with visual updates. Most time was spent with the nursing director and the nurses who were almost all inside the half who saw what was happening. Well supervised regarding the statistical design's findings, the nurses stayed with the interventions. Of course if insufficient momentum had remained in management, the effort would have been over and stopped. No one does what's already been declared a failure from the top.

Full care management returned by Week 22 (early June of that year). The run of 10 weeks below the previous average confirms the improvement. It grew to about a 17% reduction by the time Figure 1.12 ends but it was still decreasing as the client engagement on that project ended.

The test for control here is eight or more points in a row on either side of the prior average. That's obviously unlikely to occur naturally, as eight coin tosses rarely all landing heads would confirm. It's appropriate to recalculate and redraw the limits and centerline at that point, after the improvement settles into a new horizontal pattern. This underlines the importance of not resetting the limits for administrative reasons. If someone had reset the limits (say) at the end of the test or at its implementation, or as the new nurses were hired, the statistical signaling would have been lost and a very different (wrong) story would have emerged. The rules are rigorous like this so the theory can do its job.

More scientific work at Week 22 was useful. Figure 1.13 shows a clear "14-year flood" then. This one was especially easy for the nursing supervisor to understand. Because the 60,000+ members with a select set of chronic diseases were split into levels of severity, only the most severe few thousand cases received full care (Figure 1.12). A few more thousand received limited care (Figure 1.13). The remaining large number of members received almost no care (Figure 1.14). Only the most severely ill patients were in the statistical design therefore it wouldn't have seemed Figure 1.13 should show this temporary improvement. Recalling, however, the opening description of the care program, patients moved into full care and out of it as their health improved, using a set of clinical measurements and criteria. So, at periodic intervals, patients were re-leveled and as many as hundreds at a time would move from Figure 1.12 to 1.13. So the drop at Week 22 is the influx of fitter patients coming out of the statistical design, still enjoying its benefits for awhile longer. Then their admit rate creeps back up to where it had been. This creeping is because patients coming out of statistically designed care do not suddenly decide to jump into the hospital. Even if they immediately drop what they'd been hearing from their care nurse, worsening of health would not be instant. (The real situation was more complex with patients moving among all three levels of care and with some other complexities thrown in for completely new members and some other details. In fact several cohorts were used to track all of that complexity, boiled down to the simpler view here.)

Figures 1.13 and 1.14 give an additional check on the improvement inasmuch as neither reduces in admit rate overall whereas Figure 1.12 does. This is a weaker control and proof of improvement than Figure 1.12 alone using statistical control, but it's still helpful and confirmatory.

The care management program had been competitive among similar programs, but not advancing on the competition until about Week 23 (with later hindsight, rather like bottoms of recessions are not clear until later). Although patient health was the primary outcome, it's also easy to see the economic advantage promised in the original theory [2]. Conservatively (because the improvement was still growing) the 17% improvement by Week 32 on a prior admit rate of (say) 1,450 per thousand per year for the 5,000–6,000 members in full care is worth (using 5,000 members):

1,450 × 5 (thousand) × 0.17 = 1,233 fewer admits per year.

At about $10,000 average cost, this roughly estimates a healthy return, consistent with the theory. This case caused interest among competitors, who could not copy it but sent technical people to learn the methodology for their own, entirely different, solutions.

3.7 FALSE ALARM RATE IS NEITHER KNOWN NOR USEFUL IN STATISTICAL CONTROL

With a bell-shaped curve, the chance of exceeding the three standard deviation limits is 0.27%. If someone were to ask, however, whether a point outside limits could just be due to pure chance, the answer is that the question's wrong.

Figure 3.2 explained how statistical control works. The two problems, with the question as put, are first that the 0.27% number only applies for a process in perfect statistical control. That this does not happen in the real world was one of the more remarkable insights in the original work [2]. Yet it's very easy to see (Figure 3.1) with no technical knowledge.

Suppose we try some other world where processes do behave in this unrealistically controlled fashion; the answer then becomes "No." Nothing is due to pure chance although many things (such as all points inside limits) are not economic to pursue. Certainly the cause systems subtending any particular data point could be argued as triggering "by chance" but even this breaks down immediately under cross-examination. If a single data point breaks limits then its causes are by definition under statistical control theory, real, and worth making good use of. Thus, this "by chance" business has no meaning when using statistical control.

Where the question can be yielded to a small amount is that if a point outside limits can't be figured out then it's just dropped and moved on from, unexplained. It then remains a mystery. Not every mystery is worth a sledgehammer to crack because there'll be more coming and easier. So there's no meaning to the question and no usable action even if there were. Even making that series of stretches would not in any way leave a problem such as might obscure improvement. To the extent it could obfuscate improvement as a distraction, the clear answer should be helpful. This leads to profound implications of the theory of statistical control in making businesses most profitable. A necessary digression to introduce some new terms comes first, which makes the remaining discussion easier to follow.

3.8 STATISTICAL CONTROL TERMINOLOGY

Figure 3.2 provides the clearest understanding of statistical control. Common terminology also used is self-explanatory in Figure 3.3.

3.9 STATISTICS BREAKS DOWN IN UNSTABLE PROCESSES

In statistical texts and courses there are important underlying assumptions that allow statistical methods to give true statements and pictures

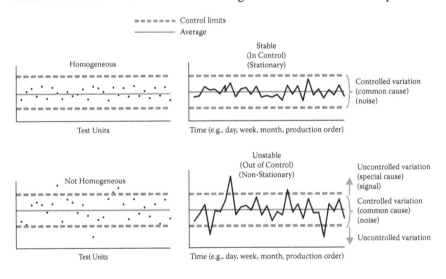

FIGURE 3.3
Terminology of statistical control.

about the real world. Industrial processes are usually unstable at the outset so statistical theory breaks down, even in the simple comparison of two averages or a pre-post in time to see if a process change helped. Fortunately the easy and correct adjustment using statistical control theory allows statistical methods to work properly.

There is a distinction between quantifying what just happened (e.g., the economy was up and down like a yo-yo but overall it netted out to a loss of $3 trillion in shareholder value) and improving something (e.g., sales had been highly volatile in the past two quarters but using statistical control it was possible to find a reliable strategy and keep such and such a percentage of our market). In the former a highly unstable phase was quantified at its end, warts and all, whereas in the latter, sense needed to be made of an equally unstable phase and turned into useful predictive information about what management actions would work. It is the controlled variation that's predictive. In the latter case, statistical control artificially excluded uncontrolled variation (outside limits) to find the inherent state of control, then used it. More than statistical control would be needed given the urgency, but it sets up a working stability as a solid foundation. In the former case, the data were correctly taken as they were found, uncensored and summarized, warts and all.

The two hypothetical cases (reflective of all real ones) are known as *enumerative* and *analytic* problems, respectively [2,11,12]. Most industrial problems are analytic. Statistical control is essential to solving industrial problems most efficiently, or sometimes at all.

It's not of course that statistical theory breaks down with an unstable process, but rather its assumptions are violated. A phrase that appears often in statistical texts is usually something like: "Assume so-many observations are drawn at random from a so-and-so distribution and are independently, identically distributed (iid)."

It's this "iid" business that's violated by unstable processes. Data outside statistical limits are tending away from iid. They come from somewhere else: of cause that's easy to find and fix and so on. So the simple adjustment ends up being the exclusion of uncontrolled variation from the calculation of average and control limits. The scientific work then proceeds, following the rules of statistical control. The examples below clarify the practical settings. All statistical control in this book follows the adjustment rules, including within statistical design analyses.

Although statistical control has been around since 1931, even today it's often someone feels the limits are "too wide" and that a tighter standard (than give or take three standard deviations) might be "more diligent." Certainly tighter limits would cause more work, but less improvement. It is easy to see why. Such tighter limits would form looser standards, not tighter ones. In the phase of using points outside limits as clues for improvement, easily fathomed when using the correct limits, people would quickly tire of what they would find to be wild goose chases. Adding meaningless work isn't a tighter standard. Then, when improvement was emerging, the tighter limits would declare victory speciously or too soon. Improvement would be reduced by relaxing its standards in this way.

Exercise 10

How would this argument be extended to Type II error (i.e., the chance of missing something important)?

3.10 ECONOMIC LOSS WITHOUT STATISTICAL CONTROL

An early example of economic loss was prevented when the request for "perfect" data in the care management case was advised against and wisely dropped. What would have happened at first would have been excessive resource consumption to make progress. That would have thrown the measurement error back out of statistical control. It's hard to give real examples because it's almost always prevented but in the rare cases where it was attempted, what happens is the measurement error gradually increases until it explodes and the measurement becomes unusable again. This can cause consternation when diligent data cleaners point to actual errors that were found and removed. The point is that without preventing them at the source, they recur. The actions taken in the care management case were sparing and, by design, attacked only the uncontrolled variation revealed by statistical control. Then each finding was developed into a system change so recurrences were prevented.

The phenomenon of throwing the thing back out of statistical control can be hard to grasp when tangible errors were found and lots of them removed. It can, however, be the ones missed that throws it back out. Some jigsaw puzzlers start with the edges to build the perimeter first. A sample of 12 statisticians and researchers in October 2013 found that, by their recollections, they always had to go through 1,000 pieces two to three times to find the last edge piece. Recollections went as high as five to six times. By analogy, measurement "perfection" is impossible and not even necessary. It just needs to be in statistical control and less than about a quarter of the total process variation. Numerical examples and details follow in Chapter 4, devoted to measurement error and solutions.

Some more cases in uneconomic actions follow, but the main point being made is the general one, of similar potential damage in all of business.

3.11 COST EXPLOSION STORY UNEXPLODED

Figure 3.4's cost control data show a problem allegedly developing. There were meetings to find an immediate solution and get costs back in line. It was felt the "problem" was suddenly "out of control" (which has a very different colloquial meaning than the statistical one, often opposite). A system change had been made in the month of July in the previous year and the consensus was that this had failed, therefore it needed to be stopped with more effective solutions implemented quickly. Someone had run the arithmetic of a statistical test by rote to "prove" the cost degradation significant.

In fact costs are largely in statistical control and a correct control chart (Figure 3.5) revealed a long-term upward trend that had been missed in a shorter-term view of the data. That trend was found by simple regression in a few seconds. The same regression found Decembers low then Januarys high. There was in reality nothing of concern in the most recent three

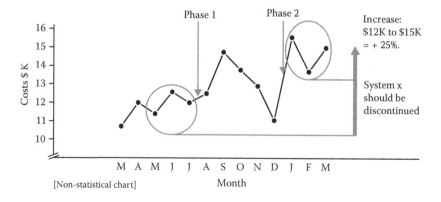

FIGURE 3.4
A problem with cost control?

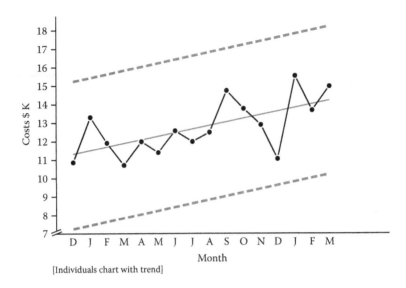

[Individuals chart with trend]

FIGURE 3.5
Initial control chart on costs.

months, with cost data falling right in line with the trend and the natural variation about it. Finding the trend is part of finding the inherent state of statistical control that is hidden in every industrial process.

By now, the momentum to remove the new July system had built and there was not much interest in Figure 3.5. This system removal was a fairly major decision though, therefore it took awhile to move through for approval. In the meantime, the statistical control chart was maintained and over the next five months emerged in Figure 3.6.

On pursuing the science underneath, there was some knowledge in the trenches as to why the December–January bump always happened. This wasn't directly germane but what was most interesting was people were asked how long the July system might take to start working. There were fairly long lag times in how costs worked after a change was made, rather like inventory levels take awhile to respond to a policy change. Without leading the witnesses, opinions varied from two to four months. It was unanimous that immediate cost reduction would be impossible. These estimates were not timed from the July Phase 1 but from the December Phase 2 when the system went live. That would put expected improvement around February to April of the following year. The statistical control limit was not broken until August but, with that hindsight, the cost reduction appeared to have started earlier. Now in August, this represented that "62-year flood" (more

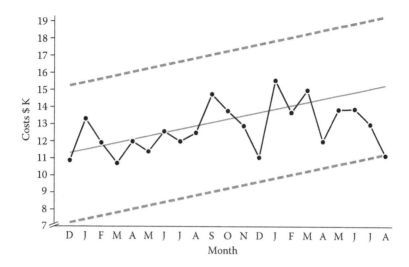

FIGURE 3.6
Emerging control chart on costs.

rare than for the care management case because those were weekly data and these monthly). Such an unprecedented reduction is next to impossible, naturally. Adding fuel to that argument, with hindsight, a few months form a downward curve leading to the eventual limit break; it's not so sudden or a single data point per se. Strictly speaking, without a designed trial, the system cannot be proven causal. But it was a good bet and one everyone found plausible, with no other possible reasons forthcoming. The long-term trend up, once revealed in this way, was easier to link to cause and had to do with complex changes in the cost structure of the business. That became the real issue needing immediate attention by engineering.

This case fragment underlines how badly statistical distribution theory breaks down when its assumptions are not so, unless adjusted correctly by statistical control in a couple of minutes' work. Notice how the first chart, which appears to be perfectly reasonable and could not have been manipulated, is in fact a gross illusion of what's really going on. The eye is seriously misled and any management conclusions distorted, really impossible. The original chart was accidental without knowledge of statistical control. No one meant to misrepresent the data.

The story didn't have a happy ending. The momentum built by the non-statistical chart and surrounding discussion remained strong. The decision to dismantle the July system had been made and pragmatically, the contradictory science (and a correct cost solution) could be reintroduced the following year after the thing had blown over. Efforts also began to arrange short courses in statistical design and control for managers. It's impossible to see these solutions without classical methods that have been available for decades. The following comes from later that year.

Exercise 11

The control charts in Figure 3.7 show production in three separate production operations. If asked to analyze the reasons this unprecedented production drop problem in all plants occurred at once in the final month shown, for what data and resources would you ask?

Exercise 12

If there might be economic loss impending, how would that soon look (in a nontechnical sentence)?

In the actual case, all three operations are in statistical control. This is not usual but it can occur. The efforts of the three plant managers and their teams were keeping production steady and presumably keeping pace with the competition. Conversely there's no improvement over the months. To the extent production was lower than market leader benchmarks, this precipitous drop was regarded as a problem. In that context, being in statistical control means the problem is chronic (i.e., will continue long term).

Statistical control theory finds the reasons for the productivity drop in all plants will be uneconomic to unravel, except by more sophisticated methods. Whether each plant increases or decreases its production each month over the last is akin to a coin toss. It's up and down about equally and this can be seen by counting the ups and downs on each chart. The three-plant drop is just like tossing three coins at once and all landing on heads. Plant productivities will all three drop every eight months in the long run. This is easy to check with three coins repeatedly tossed. They'll land all three on heads about once in every eight. Tails will do the same thing.

It might appear we're trying to have it both ways here; having argued variation within limits is not due to chance, the coin toss analogy appears to be treating it that way. The distinction is that single data points move up or down due to causes too complex, small, or many to determine easily.

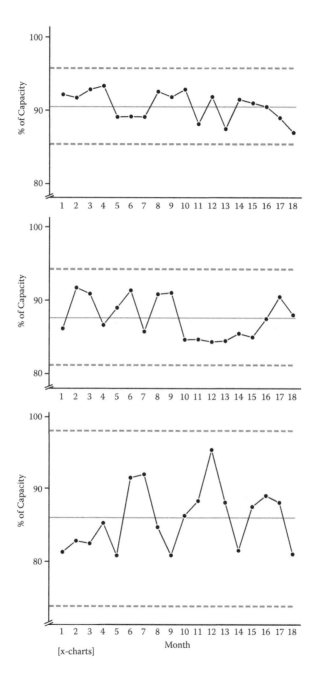

[x-charts]

FIGURE 3.7
A problem with production?

Each month took a lot of hard work to maintain a fairly constant performance. Over the longer run, however, once the probability of increasing or decreasing is found to be about a half, the coin toss analogy can be used predictively, retrospectively, to explain the phenomenon of all three dropping coincidentally.

The company was advised to run statistical designs in each plant to increase production, and also to stop the analytical efforts to find out why all three plants had dropped. Those designs appear in Chapter 6.

3.12 TESTS FOR STATISTICAL CONTROL

The primary test for control (or lack/loss of it) is the control limits. There is a variety of additional tests for control that look for trends and patterns. A powerful combination is to add just the following:

1. Run of eight or more, above or below the average line
2. Number of times average line is crossed [9]
3. Trends up or down

All four tests for control are depicted in Figure 3.8. Trends up and down are easily checked for significance with simple linear regression in a spreadsheet or statistical software.

The calculations for control limits [9] are summarized in the Method 5 box, with simplified cases most useful with statistical designs.

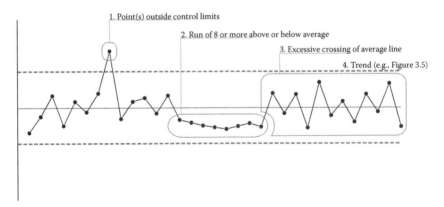

FIGURE 3.8
Four tests for control.

METHOD 5: SIMPLIFIED CALCULATION OF STATISTICAL CONTROL LIMITS

Calculations <u>exclude</u> uncontrolled variation outside trial limits, which may take more than one iteration. All chart limits at average ± 3σ̂ except range charts noted, where σ̂ is for what's <u>plotted</u>.

Pairs of charts monitor performance and its consistency separately for variables.

For p, c, and u-charts, average sums numerators then divides by summed denominators; n is the number of items per plotted point; limits change at each point.

The convention is to join the plotted points only if in production order.

"Pronunciations" and notation are given.

	Data	Chart	Average	σ̂	Notes
1.	Variable (e.g., Sales, cycle time)	"individuals" X	$\bar{X}$	$\overline{mR}/1.128$	
	mR Day: 1 2 ...	"moving range"	$\overline{mR}$		Limits at 3.27 $\overline{mR}$ and 0
2.	Variable	"X-bar" $\bar{X}(n=2)$	$\bar{\bar{X}}$	$\bar{R}/1.595$	
	$\bar{X} \rightarrow$ R Day: 1 ...	"range" R (n = 2)	$\bar{R}$		Limits at 3.27 $\bar{R}$ and 0
3.	Attribute (e.g., Proportion of leads sold)	"proportions" p	$\bar{p}$	$\sqrt{\dfrac{\bar{p}(1-\bar{p})}{n}}$	n= sample per point plotted
4.	Attribute (e.g., Number of power outages)	"counts" c	$\bar{c}$	$\sqrt{\bar{c}}$	
5.	Attribute (e.g., Admit rate)	"rates" u	$\bar{u}$	$\sqrt{\bar{u}/n}$	

Note: The constant 1.595 is derived from: $\bar{R}/1.128\sqrt{2} = \bar{R}/1.595$.

c, and numerators for p and u should be at least 5 per datum, in practice as low as about 3 is still effective. For attributes, if n is large, use X, to avoid control limits becoming oversensitive and excluding controlled variation as well as uncontrolled.

The range based estimates of σ are easier to adjust for uncontrolled variation than the usual formula.

Dropping a data set into software to give σ̂ will not give statistical control limits correctly since the variation <u>within</u> each point plotted is used as a "yardstick" for excessive variation <u>among</u> them.

3.13 STATISTICAL CONTROL INTEGRATED WITH STATISTICAL DESIGN

When integrated with statistical design, the following recaps the main applications of statistical control, worded a little better using the new language:

- To establish a working stability over time giving a clear baseline yardstick and to reveal initial improvements via uncontrolled variation (outside limits)
- To establish homogeneity among test units (e.g., nurses, retail stores) before a statistical design
- To manage implementation of a statistical design's findings and give a method to close gaps
- To support continuous improvement long term and sustain improvements

3.14 MANAGING STATISTICAL CONTROL SCHEMES

Control schemes start with what's important to the customer. Revenues, sales, delivery cycle time, and retention may be good starts. Operational issues such as production cycle time, first-time call resolution (in a customer service call center), cost per unit or transaction, and expensive errors are all balanced, because if it's being sold it may as well be made well and at the lowest cost. The general strategy is to start at the customer and work back to the supplier chain, new product/service design, and staff support.

Every new control chart has to be followed by improvement within a quarter or two, sustained long term with continuous improvement added often along the way. Accordingly, control schemes are punctuated by statistical designs. The two methods (statistical design and control) are integrated and neither works as well without the other.

Control charts start on a couple of large things and only expand to other things after the first are improved. Because they're self-teaching devices it's damaging to chart everything at once from the outset. It's impossible to improve everything at once until a start has been made with resounding breakthroughs on a couple of large things that bring clear benefit to customers and profit. The largeness has to be enough to interest executives. If lots of things are charted, it self-teaches remarkably fast that they aren't being improved and this quickly morphs to that they can't and won't be.

3.15 MECHANICS OF STATISTICAL CONTROL

The original text [2] is among the more demanding in the technical literature so not recommended for users, while remaining essential for technical leaders. Duplicating the calculation methods and details is not the purpose of this book but the Method 5 box gives more understanding of the statistical control examples used in the cases and in fact would allow a start to be made.

Figures 3.9 to 3.11 reproduce the data and calculations for the homogeneity checks by statistical control charts used in the care management and retail cases.

The larger picture is the main one, however: that statistical design provides a practical, living method for understanding how things work and how to make them work better. Distinct from many analytic techniques, it unfolds as the real world does.

3.16 WHERE DID STATISTICAL CONTROL ORIGINATE?

In 1924, Mr. R. L. Jones supervised a technical department in a research facility and had been given a specific problem to solve, assigning it to a physicist who then wrote back starting the memorandum with: "A few days ago,[*] you mentioned some of the problems associated with the development of an acceptable form of inspection report which might be modified from time to time, in order to give at a glance the greatest amount of accurate information." Statistical control prototypes were attached.

The memorandum eventually led to the publication of the landmark text [2] seven years later. It uses physics, mathematics, economics, and statistics to develop the theory. The theory was developed for anyone to use and is not restricted to manufacturing.

The text has been notoriously hard to grasp. This chapter translates its salient points into a practical user guide, without doing justice to the scope and depth of the original work as an important reference for technical leaders.

---

* [This suggests the theory was developed rather quickly.]

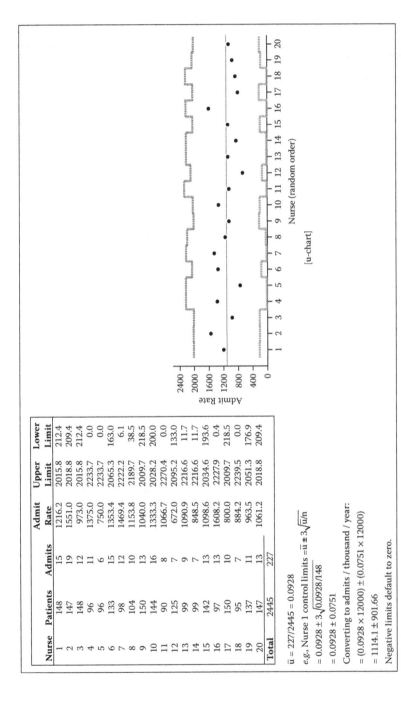

Nurse	Patients	Admits	Admit Rate	Upper Limit	Lower Limit
1	148	15	1216.2	2015.8	212.4
2	147	19	1551.0	2018.8	209.4
3	148	12	973.0	2015.8	212.4
4	96	11	1375.0	2233.7	0.0
5	96	6	750.0	2233.7	0.0
6	133	15	1353.4	2065.3	163.0
7	98	12	1469.4	2222.2	6.1
8	104	10	1153.8	2189.7	38.5
9	150	13	1040.0	2009.7	218.5
10	144	16	1333.3	2028.2	200.0
11	90	8	1066.7	2270.4	0.0
12	125	7	672.0	2095.2	133.0
13	99	9	1090.9	2216.6	11.7
14	99	7	848.5	2216.6	11.7
15	142	13	1098.6	2034.6	193.6
16	97	13	1608.2	2227.9	0.4
17	150	10	800.0	2009.7	218.5
18	95	7	884.2	2239.5	0.0
19	137	11	963.5	2051.3	176.9
20	147	13	1061.2	2018.8	209.4
Total	2445	227			

$\bar{u} = 227/2445 = 0.0928$

e.g., Nurse 1 control limits $= \bar{u} \pm 3\sqrt{\bar{u}/n}$

$= 0.0928 \pm 3\sqrt{0.0928/148}$

$= 0.0928 \pm 0.0751$

Converting to admits / thousand / year:

$= (0.0928 \times 12000) \pm (0.0751 \times 12000)$

$= 1114.1 \pm 901.66$

Negative limits default to zero.

FIGURE 3.9

Homogeneity among nurses in the care management case.

Nurse	Patients	Admits	Admit Rate	mR
1	148	15	1216.2	-
2	147	19	1551.0	334.8
3	148	12	973.0	578.0
4	96	11	1375.0	402.0
5	96	6	750.0	625.0
6	133	15	1353.4	603.4
7	98	12	1469.4	116.0
8	104	10	1153.8	315.5
9	150	13	1040.0	113.8
10	144	16	1333.3	293.3
11	90	8	1066.7	266.7
12	125	7	672.0	394.7
13	99	9	1090.9	418.9
14	99	7	848.5	242.4
15	142	13	1098.6	250.1
16	97	13	1608.2	509.7
17	150	10	800.0	808.2
18	95	7	884.2	84.2
19	137	11	963.5	79.3
20	147	13	1061.2	97.7
		Average	1115.45	343.89
		Upper limit on mR	3.27 $\overline{mR}$ =1124.52	

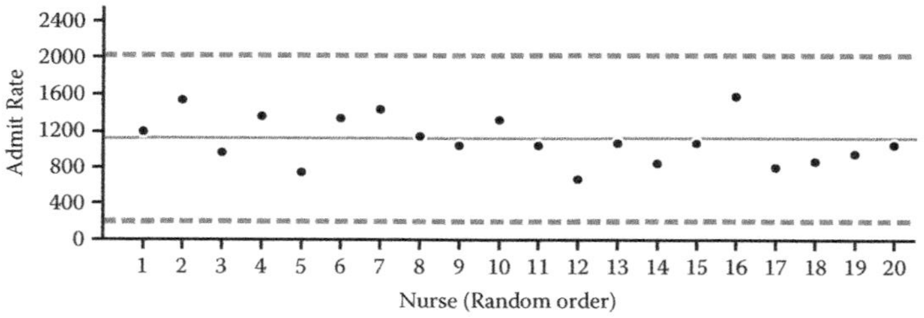

Moving range (mR) are successive differences between nurse admit rates. None exceed upper limit at 1124.52.

$\hat{\sigma} = \overline{mR}/1.128 = 343.89/1.128 = 304.87$

Control limits = 1115.45 ± 3(304.87)

i.e., 200.8 and 2030.1

FIGURE 3.10

X-chart: approximate homogeneity check among nurses.

	Previous Year's Sales		
Store	Units	Log (Units)	mR
1	56	1.748	-
2	122	2.086	0.338
3	49	1.690	0.396
4	197	2.294	0.604
5	193	2.286	0.009
6	67	1.826	0.459
7	178	2.250	0.424
8	57	1.756	0.495
9	120	2.079	0.323
10	72	1.857	0.222
11	53	1.724	0.133
12	95	1.978	0.253
13	213	2.328	0.351
14	106	2.025	0.303
15	33	1.519	0.507
16	72	1.857	0.339
17	24	1.380	0.477
18	28	1.447	0.067
19	64	1.806	0.359
20	46	1.663	0.143
21	73	1.863	0.201
22	43	1.633	0.230
23	198	2.297	0.663
24	116	2.064	0.232
25	76	1.881	0.184
26	210	2.322	0.441
27	101	2.004	0.318
28	65	1.813	0.191
29	859	2.934	1.121
30	383	2.583	0.351
31	283	2.452	0.131
32	114	2.057	0.395
Average		1.985	0.344
Upper limit on mR	$= 3.27 \times 0.344 = 1.125$		
$\hat{\sigma}$	$= 0.344/1.128 = 0.305$		
Control Limits	$=1.985 \pm 3(0.305)$		
Trial upper limit	$= 2.899$		
Excluding #29: adjusted average	$= 1.954$		
Adjusted limits	$=1.954 \pm 3(0.305)$		
	ie. 2.869 and 1.039		
Translating back:	$10^{1.039}=10.9$		
	$10^{2.869}=739.2$		

FIGURE 3.11

Homogeneity among retail stores.

4

Measurement Error and Control

Of what value is the theory of control if the observed data...are bad?

Walter Shewhart[*]

4.1 ALL MEASUREMENT SYSTEMS ARE INHERENTLY FLAWED

All measurements have error over time and when broken out in test units for statistical design. "Error" in this context does not necessarily mean someone made a mistake; it means there's a difference between the measurement and the truth.

It's a question of degree as to whether measurements are usable for statistical design and control. Because of the randomization device and the orthogonality in statistical design, some measurements can be unusable in general but perfectly acceptable in statistical design and control. Almost always, flawed measurement systems can easily be adapted to perform adequately.

The characteristics of interest are primarily *precision* and, easier to correct, *bias*. Precision is the variation of the measurement error over a sample of data. Bias is the difference between the measured value and the true value averaged over a sample.

Measurement error adds on to pure process variation as in Figure 4.1. The standard deviations (for the pure process and the measurement error) are additive to the total standard deviation according to their squares (*variances*). The pure process cannot be measured; the total process is what's really being recorded.

---

[*] Shewhart, W.A. (1931, 1980). *Economic Control of Quality of Manufactured Product*. Reprinted 1980. Milwaukee, WI: American Society for Quality Control.

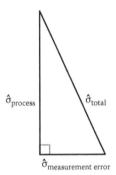

FIGURE 4.1
Observed process variation includes measurement error.

The criteria for acceptable measurement quality when using statistical design are

1. The measurement error is both stable (over time) and homogeneous (among test units).
2. The proportion of observed process variation consumed by measurement error is less than a quarter as a rule of thumb. This is calculated using:

$$\left(\hat{\sigma}_{\text{measurement error}}/\hat{\sigma}_{\text{total}}\right)^2$$

This can be assessed roughly by eye for the hypothetical example depicted in Figure 4.1. $\hat{\sigma}_{\text{measurement error}}$ is exactly half of $\hat{\sigma}_{\text{total}}$ so the proportion of total variation due to measurement error is $(1/2)^2$ = a quarter. Because of the squaring, what appears to be excessive measurement error will be passable to support a viable statistical design and control scheme.

An analogy is that of looking through a car windshield with mud accumulating in poor weather and muddy ground conditions. A thin film of mud may obscure vision somewhat but driving is still safe. Too much mud and the light is obscured, at some point making driving impossible. In process statistical design and control, "too much" is a quarter as a rule of thumb. Measurement error, if in statistical control, tends to obscure significant results in statistical design. If not in statistical control, the results and findings are liable to be wrong. The measurement error in the clinical care case is now reconstructed and compared to the criteria.

4.2 CLINICAL CARE CASE: INITIAL MEASUREMENT STUDY AND LONG-TERM CONTROLS

Measurement error was assessed by comparing admits data from two sources: authorizations and claims. Historical data were first used, going back far enough in time for the data to be mature. Rough measurement control charts were used to find spikes in specific weeks of aggregated records and these pointed directly to how their elimination could be attacked. After completion of that work, the measurement error was rechecked in Table 4.1 data for a one-month period before the test, going back just enough time for the data to be mature.

TABLE 4.1

Two Independent Measurements of Admits

(Random Allocation)	Patients	Claims	Auths.
1	148	13	15
2	147	21	19
3	148	11	12
4	96	11	11
5	96	7	6
6	133	15	15
7	98	12	12
8	104	9	10
9	150	12	13
10	144	17	16
11	90	6	8
12	125	8	7
13	99	11	9
14	99	8	7
15	142	14	13
16	97	13	13
17	150	11	10
18	95	8	7
19	137	12	11
20	147	12	13
	Average:	11.55	11.35

CALCULATION

Table 4.2 shows additional calculations needed for the measurement assessment. First, after the raw data, come the admit rates using authorizations data using the usual calculation by nurse:

[Admit count/patient] × 12,000 (scaling up from monthly to admit rate per thousand patients per year).

The next column shows the moving ranges (mR) for successive pairs of nurses. This will allow the standard deviation among nurses to be calculated.

The next column is admit rates from claims. Finally the ranges (R) between the two measurements, also their averages ($\bar{m}$) are shown.

The average hospitalizations by authorizations and claims are now close, following the earlier measurement work. Authorizations are still fewer than claims. No adjustment was made for this "fewer" bias, which will err on the pessimistic side in assessing pure measurement variation. The nurse data were arranged in random order (using the test setup), which is essential.

The lower chart in Figure 4.2 gives measurement precision by nurse by plotting the ranges between measurement sources. The upper limit uses 3.27 $\bar{R}$ as before, because there are two measurements per nurse. None of the R exceed the upper limit so measurement error is homogeneous among nurses and would support a viable test. No adjustments need to be made inasmuch as all data are within limits. The lower limit is at zero for a pair of data sources such as this.

The upper chart of $\bar{m}$ shows more than half the points plotted outside limits. This simply means the measurement $\bar{m}$ is able to discriminate among different nurse admit rates. If all points were within limits the measurement system would have no ability to discriminate even among different nurses with quite different admit rates. For this reason it is labeled a *discrimination* chart. This is a useful rough check on whether there is any discrimination at all in authorization data.

There is no need to adjust calculations for the points outside limits. In range charts this matters a lot but in the $\bar{m}$ chart (especially with six points above and below limits), it will just move the average line slightly and not

TABLE 4.2

Admit Rate Measurement Study Data

Nurse (Random Allocation)	Patients	Claims	Auths.	Admit Rate (Auths.)	mR	Admit Rate (Claims)	R	m̄
1	148	13	15	1216.2		1054.1	162.2	1135.1
2	147	21	19	1551.0	334.8	1714.3	163.3	1632.7
3	148	11	12	973.0	578.0	891.9	81.1	932.4
4	96	11	11	1375.0	402.0	1375.0	0.0	1375.0
5	96	7	6	750.0	625.0	875.0	125.0	812.5
6	133	15	15	1353.4	603.4	1353.4	0.0	1353.4
7	98	12	12	1469.4	116.0	1469.4	0.0	1469.4
8	104	9	10	1153.8	315.5	1038.5	115.4	1096.2
9	150	12	13	1040.0	113.8	960.0	80.0	1000.0
10	144	17	16	1333.3	293.3	1416.7	83.3	1375.0
11	90	6	8	1066.7	266.7	800.0	266.7	933.3
12	125	8	7	672.0	394.7	768.0	96.0	720.0
13	99	11	9	1090.9	418.9	1333.3	242.4	1212.1
14	99	8	7	848.5	242.4	969.7	121.2	909.1
15	142	14	13	1098.6	250.1	1183.1	84.5	1140.8
16	97	13	13	1608.2	509.7	1608.2	0.0	1608.2
17	150	11	10	800.0	808.2	880.0	80.0	840.0
18	95	8	7	884.2	84.2	1010.5	126.3	947.4
19	137	12	11	963.5	79.3	1051.1	87.6	1007.3
20	147	12	13	1061.2	97.7	979.6	81.6	1020.4
Average:		11.55	11.35	1115.4	343.9	1136.6	99.8	1126.0

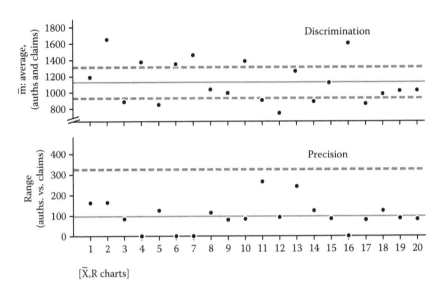

[X̄,R charts]

FIGURE 4.2
Measurement control charts by nurse.

CALCULATION (METHODS 2,5)

Using the usual statistical control calculations:

Upper limit on $R = 3.27\overline{R} = 3.27 \times 99.8 = 326.4$
$\hat{\sigma}_{measurement} = \overline{R}/1.128 = 99.8/1.128 = 88.5$
Among nurses, upper limit on $mR = 3.27\overline{mR} = 3.27 \times 343.9 = 1124.6$
$\hat{\sigma}_{total} = \overline{mR}/1.128 = 343.9/1.128 = 304.9$
Control limits on $\overline{m} = \overline{\overline{m}} \pm 3\hat{\sigma}_{\overline{m}}$
$= 1126.0 \pm 3(\overline{R}/1.595)$
$= 1126.0 \pm 3(99.8/1.595)$
$= 1126.0 \pm 187.7$
i.e., 938.3 and 1313.7

Therefore proportion of variation in total admit rate obscured by measurement error variation = $(88.5/304.9)^2 = 0.084$ (i.e., 8.4%).

change the story. The conclusion is already clear: the measurement does discriminate. The discrimination chart is one of a few occasions where the more uncontrolled the variation, the better. The cause is, of course, the changing admit rate by nurse.

Because 0.084 is less than a quarter, the measurement error is perfectly acceptable to support the test. At the outset, it hadn't been until the work to remove spikes in measurement error (by nurse and by week) was completed. It's not enough, however, to find measurement error homogeneous among nurses with just 8.4% of total variation consumed. It also has to be stable over time and kept under statistical control as in the next section.

Exercise 13

How could removing <u>all</u> actual measurement errors found possibly hurt?

4.3 ESTABLISHING A MEASUREMENT CONTROL SCHEME

Weekly data for all nurses combined follow in Table 4.3, including the usual conversion to admit rates per thousand members per year. Inasmuch as the measurement uses the conversion it's easier to work in that basis. In practice the measurement control scheme was established early and then updated weekly. The following uses the final version from Week 32.

The average admit rates in Table 4.3 exclude the last five weeks' data because claims data are clearly not yet mature, dipping below the authorizations data they ordinarily exceed. These are an administrative artifact and not relevant to the measurement error control scheme.

The bias in authorizations always being lower than mature claims data won't affect the statistical design, nor the statistical control on the admit rate used as a relative comparison week to week. If the exact data were needed for financial calculations the bias can easily be adjusted out by adding the fairly constant difference between the measurements. In assessing measurement precision over time, however, leaving the bias in will give a pessimistic assessment of precision. The bias is artificially adjusted out accordingly. Because the bias in authorizations admit rate with respect to claims admit rate is –183, it's subtracted out in Table 4.4.

TABLE 4.3

Measurement Error Data over Time

Week	Patients	Admits Auths.	Admits Claims	Admit Rate Auths.	Admit Rate Claims
1	3302	90	106	1417.3	1669.3
2	3302	117	133	1842.5	2094.5
3	3302	106	124	1669.3	1952.8
4	3302	91	103	1433.1	1622.0
5	3302	99	113	1559.1	1779.5
6	3363	93	106	1438.0	1639.0
7	3363	84	93	1298.8	1438.0
8	3363	100	111	1546.2	1716.3
9	3363	114	121	1762.7	1870.9
10	3410	90	99	1372.4	1509.7
11	3410	107	119	1631.7	1814.7
12	3410	103	107	1570.7	1631.7
13	3410	86	96	1311.4	1463.9
14	3428	98	111	1486.6	1683.8
15	3428	68	85	1031.5	1289.4
16	3428	80	100	1213.5	1516.9
17	3428	92	105	1395.6	1592.8
18	5238	186	212	1846.5	2104.6
19	5238	166	186	1648.0	1846.5
20	5238	152	180	1509.0	1786.9
21	5238	178	196	1767.1	1945.8
22	5238	151	165	1499.0	1638.0
23	5663	150	160	1377.4	1469.2
24	5663	153	163	1404.9	1496.7
25	5663	152	156	1395.7	1432.5
26	5663	147	166	1349.8	1524.3
27	5192	132	151	1322.0	1512.3
28	5192	136	127	1362.1	1272.0
29	5192	143	127	1432.2	1272.0
30	5192	132	119	1322.0	1191.8
31	5043	121	98	1247.7	1010.5
32	5043	116	38	1196.1	391.8
Average (Weeks 1–27)				1485.2	1668.2

Auths. bias

= 1485.2 – 1668.2   (w/o last 5)

= –183.0

TABLE 4.4

Bias Adjusted Measurement Error

Week	Patients	Admits		Admit Rate		Bias Adjusted		R	m̄	mR
		Auths.	Claims	Auths.	Claims	Auths. (m_1)	Claims (m_2)			(auths.)
1	3302	90	106	1417.3	1669.3	1600.4	1669.3	68.9	1634.8	
2	3302	117	133	1842.5	2094.5	2025.6	2094.5	68.9	2060.0	425.2
3	3302	106	124	1669.3	1952.8	1852.3	1952.8	100.4	1902.5	173.2
4	3302	91	103	1433.1	1622.0	1616.1	1622.0	5.9	1619.1	236.2
5	3302	99	113	1559.1	1779.5	1742.1	1779.5	37.4	1760.8	126.0
6	3363	93	106	1438.0	1639.0	1621.0	1639.0	18.0	1630.0	121.1
7	3363	84	93	1298.8	1438.0	1481.9	1438.0	43.9	1459.9	139.2
8	3363	100	111	1546.2	1716.3	1729.3	1716.3	13.0	1722.8	247.4
9	3363	114	121	1762.7	1870.9	1945.8	1870.9	74.8	1908.4	216.5
10	3410	90	99	1372.4	1509.7	1555.5	1509.7	45.8	1532.6	390.3
11	3410	107	119	1631.7	1814.7	1814.7	1814.7	0.1	1814.7	259.2
12	3410	103	107	1570.7	1631.7	1753.7	1631.7	122.0	1692.7	61.0
13	3410	86	96	1311.4	1463.9	1494.5	1463.9	30.6	1479.2	259.2
14	3428	98	111	1486.6	1683.8	1669.6	1683.8	14.2	1676.7	175.1
15	3428	68	85	1031.5	1289.4	1214.5	1289.4	74.8	1252.0	455.1

Continued

TABLE 4.4 (Continued)

Bias Adjusted Measurement Error

Week	Patients	Admits		Admit Rate		Bias Adjusted		R	$\overline{m}$	mR (auths.)
		Auths.	Claims	Auths.	Claims	Auths. (m_1)	Claims (m_2)			
16	3428	80	100	1213.5	1516.9	1396.6	1516.9	120.3	1456.7	182.0
17	3428	92	105	1395.6	1592.8	1578.6	1592.8	14.2	1585.7	182.0
18	5238	186	212	1846.5	2104.6	2029.5	2104.6	75.1	2067.1	450.9
19	5238	166	186	1648.0	1846.5	1831.0	1846.5	15.5	1838.8	198.5
20	5238	152	180	1509.0	1786.9	1692.0	1786.9	94.9	1739.5	139.0
21	5238	178	196	1767.1	1945.8	1950.1	1945.8	4.3	1948.0	258.1
22	5238	151	165	1499.0	1638.0	1682.1	1638.0	44.1	1660.1	268.0
23	5663	150	160	1377.4	1469.2	1560.4	1469.2	91.2	1514.8	Improvement
24	5663	153	163	1404.9	1496.7	1588.0	1496.7	91.2	1542.3	starts
25	5663	152	156	1395.7	1432.5	1578.8	1432.5	146.3	1505.6	
26	5663	147	166	1349.8	1524.3	1532.9	1524.3	8.6	1528.6	
27	5192	132	151	1322.0	1512.3	1505.1	1512.3	7.2	1508.7	
28	5192	136	127	1362.1	1272.0	1545.1	1272.0	273.2	1408.5	
29	5192	143	127	1432.2	1272.0	1615.2	1272.0	343.3	1443.6	
30	5192	132	119	1322.0	1191.8	1505.1	1191.8	313.2	1348.5	
31	5043	121	98	1247.7	1010.5	1430.7	1010.5	420.2	1220.6	
32	5043	116	38	1196.1	391.8	1379.2	391.8	987.3	885.5	
Average (Weeks 1–27)				1485.2	1668.2	1668.2	1668.2	236.4	(weeks 1–22):	236.4

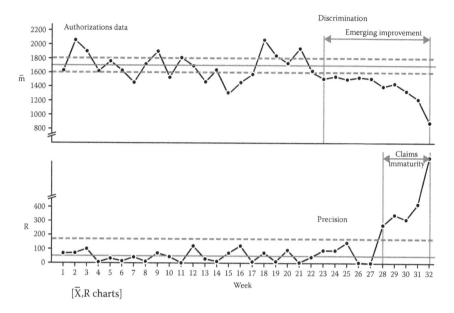

FIGURE 4.3
Measurement control scheme over time.

The two data sources now match in their averages over the first 27 weeks, as shown. The ranges and the average for each pair of measurements, weekly, is also shown. The final column gives the moving ranges for adjacent week admit rates, using authorizations.

Figure 4.3 shows the discrimination and precision charts as before, but over the 32 weeks, for all nurses together. The measurement control scheme continually provided useful information. From a viewpoint at Week 32, the scheme had prevented all questions about measurement. Any that were asked could be readily answered with a glance at the chart each week.

Starting, as always, with the range chart, it illustrates measurement precision and that the measurement error had remained stable each week until the most recent five weeks due to claims data being immature. Over that lag, authorizations data were used exclusively to track the emerging improvement and troubleshoot it, with the increasing confidence over the weeks that it would be found free of serious measurement error that could obscure the reality. The ranges move out of statistical control over that past five weeks, exploding off the chart in the final one. This gave a useful rule of thumb as to how soon data from claims could be used: on this scale it's up to Week 27 from Week 32's viewpoint.

CALCULATION (METHODS 2,5)

Measurement precision:

$\overline{R}$ = 117.8 (Weeks 1-32)
Trial upper limit on R = $3.27\overline{R}$ = 3.27 × 117.8 = 385.2
Removing #31 and #32:
Adjusted $\overline{R}'$ = 78.71
2nd. trial upper limit = $3.27\overline{R}'$ = 3.27 × 78.71 = 257.4
Removing #28, #29 and #30:
Re-adjusted $\overline{R}''$ = 53.03
3rd. trial upper limit = $3.27\overline{R}''$ = 3.27 × 53.03 = 173.4
No further R exceeds 173.4, confirming the eyeball guess to remove
the last 5 weeks.
So, upper limit on R chart = 173.4
$\hat{\sigma}_{measurement} = \overline{R}''/1.128$ = 53.03/1.128 = 47.0

Measurement discrimination:

$\hat{\sigma}_{\overline{m}} = \overline{R}''/1.595$ = 53.03/1.595 = 33.25
$\overline{\overline{m}}$ (Weeks 1-22, before process shift) = 1701.9
Control limits on $\overline{m}$ chart = $\overline{\overline{m}} \pm 3\hat{\sigma}_{\overline{m}}$
= 1701.9 ± 3(33.25)
= 1701.9 ± 99.75
i.e. 1602.2 and 1801.7

Total process:

$\overline{mR}$ = 236.4 (auths. from weeks 1-22 before process shift)
Upper limit on mR = $3.27\overline{mR}$ = 3.27 × 236.4 = 773.0
No mR exclusions.
$\hat{\sigma}_{total} = \overline{mR}/1.128$ = 236.4/1.128 = 209.6

The discrimination chart shows about half the measurements outside limits, indicating there is discrimination in the measurement system $\overline{m}$. The implementation improvement becomes clear starting in Week 23 and the variation from week to week clearly tightens. This improvement continues to emerge (as Figure 1.12 showed, free of claims data immaturity, being entirely authorizations based) but is artificially exaggerated by the

CALCULATION (METHOD 2)

The box after Figure 4.3 gives the iterative calculations to remove uncontrolled variation in measurement error. This gives:

$$\hat{\sigma}_{\text{measurement error}} = 47.0$$

Also

$$\hat{\sigma}_{\text{total}} = 209.6$$

Accordingly, the proportion measurement variation = $(47.0/209.6)^2$ = 0.05 (i.e., 5%).

claims immaturity after Week 27. This isn't a problem given the purpose of the chart. The sudden change in the slope of the line at Week 28 is for the same reason of claims' immaturity. All of these inferences may be checked and confirmed in the source data of Table 4.4.

At 5% of total variation, also stable, this is all in close agreement with the earlier assessment by nurse (at 8.4%) and is completely acceptable. Certainly, the measurement error was made stable and with improved precision at the outset and then maintained by the measurement control scheme.

A useful hybrid chart is provided by superimposing the measurement control chart as a gray bar within the main statistical control chart. This even more readily answers or prevents any questions about measurement error that would otherwise be unanswerable in any objective way. If people are not confident in the measurement system (as they shouldn't be without evidence) it will thwart the improvement itself.

It's common, when dealing with surprising test findings, to suspect the measurement system and request a reanalysis after "clean-up." Of course the correct approach is to provide the measurement control chart and decline the reanalysis. It's more pragmatic, however, to complete the reanalysis and show reduction in statistical significance of results. Were the clean-up to have helped, the expectation would be for more significance.

5

Statistical Design

To find out what happens to a system when you interfere with it, you have to interfere with it...

George Box*

5.1 ADVANTAGES OF LARGE STATISTICAL DESIGN

There is a mathematical proof [13] that, given the option of replicating fewer interventions, or adding more interventions, more improvement occurs with more interventions. The advantages of large statistical design, introduced by the opening cases in care and retailing, are

1. Establishing cause and effect of each intervention directly by randomized orthogonal design.
2. A wider inductive basis, meaning that testing an intervention in the presence of many others also being varied, gives a real-world test rather than an artificially controlled one. This means the findings will more likely implement, consistent with the prediction calculated from the test. In other words the findings are more likely repeatable, without literally repeating the test.
3. When testing 20+ interventions, their total number of potential combinations exceeds a million. Large statistical design jumps approximately to the best of these. Depending on the degree to which interactions can be assessed, the approximation is correspondingly closer to optimal. In the care management case, interactions could not be assessed. In the retailing case they were.

---

* Box, G.E.P. (1966). Use and Abuse of Regression. *Technometrics*, 8(4): 625–629.

4. Finding and quantifying interactions is in itself productive, as the retailing case began to show.

5. When combined with statistical control, the importance of controlling in the real world becomes clear in both the process and its measurement error. At implementation, statistical control reveals disparities between the test conditions and the constantly shifting ones in the real world. This gives an advantage over refining or validation tests that presume a static "correct answer" within the original test rather than a dynamic one to be revealed by statistical control after the test, although well approximated in that first large test.

The care management and retail sales cases introduced the large designs at work. It is useful to next know more about how they are constructed, with more insight also into how they are analyzed. In particular, interactions need further methodology beyond the start made in the retailing case. The large designs are better understood by building up from the small designs. The small designs are called full factorials. The larger designs are of two types, called fractional factorials and multifactorials. The designs are developed, still using only simple arithmetic.

5.2 TWO-LEVEL DESIGNS

Although there are several other types of statistical design, the emphasis here is on many interventions, each at two levels (test + vs. counterfactual −) for flexibility and effectiveness in solving problems generally. More than two levels are easy to accommodate, especially three or four (e.g., pricing), but this has not been found major in improving most business processes. Sometimes a small design with a couple of interventions is right for a specific problem but generally the larger ones have been found more effective and faster.

5.3 FULL FACTORIAL DESIGNS

Figure 1.2 depicted a full factorial design in two interventions. It was common sense that it included all four possible combinations of the two interventions, each at two levels. Those four combinations were

1. Test neither +
2. Test one +
3. Test the other +
4. Test both +s

and the way that was notated was

A	B
−	−
+	−
−	+
+	+

This is called the *design matrix* and is all that's needed to run the test.

When the results are in, the design matrix is augmented to the analysis matrix by including the interaction AB, by multiplying A and B for each row using basic algebraic rules (e.g., for row 1: $- \times - = +$):

A	B	AB
−	−	+
+	−	−
−	+	−
+	+	+

Whereas Figure 1.2 depicted a plane (flat), AB puts a twist in it. This makes sense because if A and B really do interact (either synergistically or antagonistically) then the whole system would move, not just the spot where A and B are both +.

It's easy to see all four combinations with just two interventions. The pattern is alternating signs for A and alternating every other sign for B. This pattern is extended to write down the design matrix for the next full factorial design in three interventions:

A	B	C
−	−	−
+	−	−
−	+	−
+	+	−
−	−	+
+	−	+
−	+	+
+	+	+

Augmenting the analysis matrix by multiplying the columns for A, B, and C gives:

A	B	C	AB	AC	BC	ABC
−	−	−	+	+	+	−
+	−	−	−	−	+	+
−	+	−	−	+	−	+
+	+	−	+	−	−	−
−	−	+	+	−	−	+
+	−	+	−	+	−	−
−	+	+	−	−	+	−
+	+	+	+	+	+	+

The next full factorial would be in four interventions (A, B, C, and D) and 16 rows. This is rapidly going to get out of hand inasmuch as five interventions would be in 32 rows and already exceed the resources available for many business tests. Certainly, testing the 20+ interventions where statistical design is most effective would be out of the question. The fractional factorial designs are therefore needed.

5.4 FRACTIONAL FACTORIAL DESIGNS

Because interactions among more than two interventions are usually small, ABC typically wastes resources. So, the ABC column could be built into the design matrix and a new intervention (D) added, like this:

A	B	C	D
−	−	−	−
+	−	−	+
−	+	−	+
+	+	−	−
−	−	+	+
+	−	+	−
−	+	+	−
+	+	+	+

This is called a half-fraction factorial design because it uses half of the 16 rows that would have been needed for a full factorial in A, B, C, and D. A real example illustrates how it's used.

5.5 BACKPACKING CASE

Pausing to note that it is unwise to begin real industrial problem-solving with a small statistical design (for similar reasons first skydives are not attempted from 100 feet under an illusion of being safer), this next real (but small) case ran July 6–11, 1998, in the Bridger wilderness in Wyoming, at altitudes above 10,000 feet. With weight carried in backpacks carefully controlled in the packing (by outdoor experts back at base camp) there was no room for statistical books or laptops. One pencil, one sheet of paper, and mental arithmetic were used throughout. The statistical design was written at the top of the sheet then analyzed in a couple of lines after being run.

5.6 DISCOVERY

Someone had asked what happened to heart rate at altitude, especially with 60-pound backpacks and long hikes by middle-aged people. In discussion, the statistical design that would answer everyone's questions emerged. It tested the following interventions:

A: Subject	– #1	
	+ #2	
B: Pulse taking	– At rest	
	+ After vigorous exercise	
C: Day	– First two	
	+ Last two	
D: Pack	– Light daypack	
	+ Full backpack	

Two adults volunteered for the testing (named here as subject #1 and subject #2). Coin tosses were used to randomize.

5.7 MANAGING THE TEST

Each row of the design was verbally translated into daily procedures. For example, the first row runs within the first two days and measures #1's pulse at rest after wearing the light daypack. The last row runs within the last two days and measures #2's pulse after vigorous exercise and wearing the full backpack. The other six rows are easy to read and translate into practical procedures like this.

Standards for each intervention were agreed and remembered (e.g., how long walking fast and how long at rest before pulse taking). Instructions on how to measure and record pulse were also agreed upon. Pulse would be taken over one minute.

So, each subject was planned to conduct four of the rows, for the total of eight. This meant one measured test per day, over four days, for each subject. No time was free to establish statistical control before the test. For this reason and to establish significance of results, partial replication [12] where some of the rows are run again was designed. There was time for just four of these, selected as rows 1, 4, 5, and 8. These extra four runs were designed in advance to follow the original eight runs: two in the first two days and two more in the last two days. On days where replicates were being run, one was run in the morning and the other in the afternoon. (This also allowed a rough check on whether morning or afternoon mattered to pulse.)

It's important to define the analysis method in advance to be objective. Because no computer was available, and the analysis of partially replicated designs needs one, the plan was to analyze just the original eight results and use the partial replicates only for rough statistical control and also significance.

5.8 MEASUREMENT QUALITY

Suspicious of measurement error in the watches worn by both subjects (and in reading them), the minute was timed by each subject and the pulses right between two shouts of "Now!" used. In practice the two shouts of "Now!" were almost on top of each other so the pulse recorded was pretty clear.

Other hikers in the pack, hearing the "Now!" shouts each day, became curious and asked to see the final report on the upcoming long bus ride home. Most had thought they hated statistics, having had to take courses at some point. This they liked though, because they saw where the numbers were coming from and were interested in seeing how the puzzle would be solved. No one cared much about the mathematics inside. They just knew it was there. This is a little like enjoying reading stories but not a dictionary of words listed.

5.9 EXPLORATORY ANALYSIS

The pulse rates recorded were:

A	B	C	D	Pulse 1	Pulse 2	Range
−	−	−	−	104	104	0
+	−	−	+	60		
−	+	−	+	114		
+	+	−	−	92	96	4
−	−	+	+	98	99	1
+	−	+	−	50		
−	+	+	−	106		
+	+	+	+	58	56	2

As planned, the analysis uses only the original eight data points. The average heart rate for A+ is $(60 + 92 + 50 + 58)/4 = 65$.

Notice, as in the large case designs, that within the four rows used in the A+ average, every other intervention is found to have two + and two −. So even if B, C, or D has a large effect, it will cancel here. This is the orthogonal property at work again. It keeps all the interventions independent of each other.

For A− the average is $(104 + 114 + 98 + 106)/4 = 105.5$ and again everything else cancels.

Therefore the A+ subject had a lower heart rate than the A− by $65 − 105.5 = −40.5$ beats per minute. Similar calculations for B, C, and D give all effects, with the canceling of everything else at work in all of them:

$$A: \quad 65 − 105.5 = −40.5$$
$$B: \quad 92.5 − 78 = 14.5$$
$$C: \quad 78 − 92.5 = −14.5$$
$$D: \quad 82.5 − 88 = −5.5$$

At first glance it's a little tricky to accept that everything else cancels whenever we concentrate on one intervention. It might still appear, even after a couple of cases, that too many things are changing at once and these naïve averages are not pure effects of A, B, C, and D. In fact they are, and in this smaller example it is easy to see more about why. We can also now see increasingly clearly why the designs work, by going a little deeper into this canceling.

C is the easiest to see because it falls in two vertical chunks of 4 "–" and then 4 "+". When averaging the C+ pulses, notice what happens to B. It has 2 "–" and 2 "+" in perfect balance. So, even if B has a huge effect, it will cancel itself. We can try this by adding (say) 100 to each B+ row's pulses. C stays at –14.5 and B becomes 114.5. Even adding a million to each B+ leaves C unchanged at –14.5 (and B at 1,000,014.5). The same phenomenon occurs for all four letters. The trick of adding a million in this way helps to clarify what's happening, by trying the ridiculous, yet everything still works.

All we're seeing, again, is that the familiar notion of a (one-dimensional) balance scale, that lies beneath our belief in test versus control also works in four dimensions (this pulse rate case) but is a little easier to "see" than the 19–intervention care management case or the 8-intervention retail case.

5.10 WHAT MIGHT THE INITIAL RESULTS MEAN?

Taking the pulse results at face value initially, A– was a little concerned at his higher pulse especially because he had run five miles a day for years whereas A+ did not exercise routinely. The 14.5 higher pulse after vigorous exercise (B+) appeared to make sense and so did the –14.5 lower pulse in the last two days with acclimatization to the high altitude. That the full 60-pound backpack (D+) did not increase pulse and in fact seemed if anything to lower it slightly (–5.5) was surprising at first. Until the hikers were quizzed as to how they had hiked and they said they'd always walked as fast as they could, keeping in mind they had several miles ahead of them. Therefore they'd probably adjusted naturally for the heavy backpack and exerted themselves about the same as when wearing the light daypack, simply by walking slower. Although the reduced pulse with the heavy backpack is only slightly lower, it may be real if the strength, stamina, and psychology of carrying a 60-pound weight nets out by running the heart a little farther inside its limit.

5.11 EXPLORING INTERACTIONS

Squeezing more out of the data: all 2×2s can be tabled and graphed. This will show whether any interactions between pairs of interventions do especially well (or poorly). The retail case used heredity of effects: interventions with large effects tend to interact. Thus the first interactions to look at will be anything involving A (because it has the largest effect). Looking at the pairs: AB, AC, and AD as 2×2s gives Table 5.1.

Graphing these averages for each quadrant of each 2×2 gives Figure 5.1. Recalling that with zero interaction the lines would be parallel (Figure 5.2), because in the slope we'd see the effect of whatever's plotted on the horizontal (e.g., here, A+ slopes far lower in pulse), and in the gap we'd see the effect of whatever's plotted in the vertical split. Therefore the lines would have to be parallel.

Conversely, if one subject had overnight between Days 2 and 3, found relief from a bout of ventricular ectopy (extra beats) when exercising, we might have seen the faint dotted line on the AC graph. (Or it might have crossed the other way if the extra beats were hard to feel, giving an apparently lower pulse as an artifact, the other dotted line).

To quantify objectively whether the lines are parallel using the same mathematical rules as were used for each single intervention the diagonals on each 2×2 are averaged and subtracted as in the retail case. From Table 5.1, these diagonals are:

$$AB = (101 + 75)/2 - (110 + 55)/2 = 5.5$$
$$AC = (109 + 54)/2 - (102 + 76)/2 = -7.5$$
$$AD = (105 + 59)/2 - (106 + 71)/2 = -6.5$$

These are now also ready for significance calculations, after seeing a quicker, easier, safer way to do what was just done.

TABLE 5.1

All 2×2s Involving A

B	−	104		60		C	−	104		60		D	−	104		92	
			101		55				109		76				105		71
		98		50				114		92				106		50	
		114		92				98		50				114		60	
			110		75				102		54				106		59
	+	106		58			+	106		58			+	98		58	
		−	A	+				−	A	+				−	A	+	

Note: Italics = averages per cell.

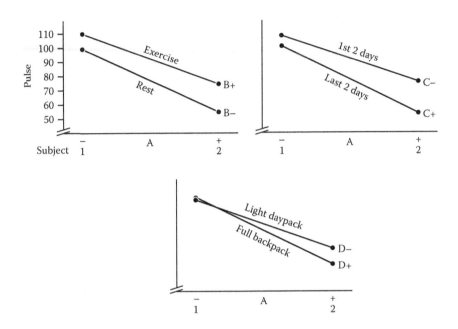

FIGURE 5.1
Graphs of all pair interactions involving A.

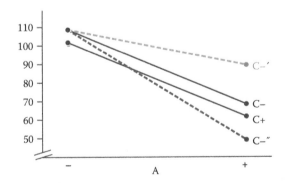

FIGURE 5.2
What zero interaction would look like.

5.12 SIMPLER ANALYSIS

Recapping all the results thus far, the original design matrix is augmented to the analysis matrix with all pairs involving A as follows:

	A	B	C	D	AB	AC	AD	Pulse
	−	−	−	−	+	+	+	104
	+	−	−	+	−	−	+	60
	−	+	−	+	−	+	−	114
	+	+	−	−	+	−	−	92
	−	−	+	+	+	−	−	98
	+	−	+	−	−	+	−	50
	−	+	+	−	−	−	+	106
	+	+	+	+	+	+	+	58
Average +	65	92.5	78	82.5	88	81.5	82	
Average −	105.5	78	92.5	88	82.5	89	88.5	
Effect	−40.5	14.5	−14.5	−5.5	5.5	−7.5	−6.5	
Grand average	= 85.25							

This gives an easier way to run the analysis already done and also have accounting checks (inasmuch as average+ and average− always must grand average 85.25). The interaction columns are again found by multiplying using the usual algebraic rules: for example, AB uses the original A and B columns, starting out:

$$
\begin{array}{ccccc}
- & \times & - & = & + \\
+ & \times & - & = & - \\
- & \times & + & = & - \\
+ & \times & + & = & +
\end{array}
$$

and so on. For now, just the pair interactions involving A are shown, under heredity of effects.

5.13 STATISTICAL SIGNIFICANCE

It turns out that everything, including the three interactions among pairs, is significant, with noise at ±3.04 (see next box). The choice of a two-sided significance test comes from the research question. No one asked: "What reduces pulse?" (one-sided) but instead: "What happens to pulse?" (two-sided).

CALCULATION (METHODS 2 AND 3)

The ranges from partial replicates are: 0, 4, 1, and 2 beats per minute, averaging $\overline{R} = 1.75$. It's a weak test for statistical control with just four partial replicates but a quick rough check finds the upper control limit on the ranges is $3.27\,\overline{R} = 3.27 \times 1.75 = 5.72$ and so pulse appears stable in this regard.

The yardstick for significance (i.e., to separate signals from noise) is given, as in the retail case, by:

$$\text{Noise} = \pm t\hat{\sigma}\sqrt{(4/N)}$$

where $\hat{\sigma} = \overline{R}/d_2 = 1.75/1.128 = 1.551$ and t is the t–statistic with $4(2-1) = 4$ degrees of freedom (i.e., the number of partial replicates).

Therefore, $t_4 = 2.776$ (from statistical tables).

N is the number of data points used to estimate intervention effects: here 8.

Therefore, noise $= \pm 2.776(1.551)\sqrt{(4/8)} = \pm 3.04$.

The reason everything is significant is that the design was loaded with interventions suspected as having strong effects on pulse. Large effects tend to interact and they all did. Interactions with A remain most likely in view of its largest magnitude, far larger than any other effect. This frivolous test finding everything significant is next to impossible in real business cases, especially with large designs.

The normal plot is ill-advised in this case because everything is significant whereas the normal plot uses the insignificant interventions as a yardstick for the significant ones. There's no danger of being misled in a real problem because if attempted, the normal plot does not come close to a straight line in its center. In this case the partial replicates provided the necessary calculations. It is good design to include a few partial replicates to double check everything is as it should be. In one case out of hundreds, about three quarters of the interventions were found significant, therefore the test was extended by replication to be sure on significance.

Exercise 14

What's the flaw in this method of assessing significance?

5.14 SOLVING THE PUZZLE

The missing piece of the puzzle was why the subjects differed so markedly in pulse. As the bus descended over the first several hours of the return journey, reaching Laramie (7,150 feet above sea level) the subjects' pulses dropped (at rest) to 74 and 48 beats per minute, respectively. The difference of 26 beats was smaller than the 40.5 in the test but still clear. In other words, the exerciser (who'd had readings above 100 at altitude) clearly dropped more than the nonexerciser suggesting he was more susceptible to altitude. Near sea level a couple of days later, the normal resting pulses were taken at 66 and 50, respectively. One subject (A+) then mentioned being on blood pressure medication. A doctor on board confirmed this would reduce pulse. As in industrial cases, there had understandably been a little initial reluctance to divulge what was behind the facts already clear in the analysis.

5.15 ALIASING

The backpacking analysis ignored, for now, interactions between the pairs CD, BD, and BC, using heredity of effects. If multiplied out, CD, BD, and BC give the same columns of + and − signs as AB, AC, and AD, respectively. So, the interaction results were really:

> AB and CD together
> AC and BD
> AD and BC

This is called the aliasing scheme (because CD is an alias for AB and so on) and is written (ignoring interactions among three and four interventions) as:

$$AB = CD$$
$$AC = BD$$
$$AD = BC$$

The interactions with A are more likely but aren't assured.

5.16 ANALYSIS OF ALL PAIR INTERACTIONS

Having seen this aliasing limitation, a closer look at the retail and care management cases shows how large designs accommodate more (and more complex) aliasing. The full analysis matrix for the retailing case is shown in Figure 5.3, including the aliasing scheme for pair interactions to the right, above. Continuing for now with the change in sales results, the aliasing scheme extends to interactions involving three or more interventions but these tend to be small and thus are excluded for simplicity. In practice, good design remains suspicious of higher-order interactions such as this and they're checked behind the scenes. In hundreds of cases, one interaction among four interventions was found and explained, but they are rare. The aliasing schemes are generated mathematically but, once done like this, are easily checked by multiplying pairs of columns.

The effects of each intervention are reproduced at the foot of Figure 5.3 and the interaction effects are also shown. These are calculated in the same way as interventions were in the original case, just averaging + versus − rows for each interaction column and their difference gives the effect. To see which interactions are significant, the same noise calculation as for the interventions is used: ±43.2, recapping.

AG is confirmed as highly significant, also AB and AC although much less so. To ensure AG was correctly diagnosed, using heredity of effects, would require another test, being a simple 2×2 in A and G: a full factorial in the two interventions. Looking at the aliasing scheme, four interactions appear in the AG column. This is written:

$$AG = CD = BH = EF$$

This does not mean their true effects are all equal, rather that the design cannot tell them apart. It's a little like walking past a room and hearing four people in furious debate but not knowing who's speaking until opening the door to see. It's liable to be one person but it could be a couple talking over each other or all four in a melee. Effects heredity tips the scales toward AG and only AG.

Salespeople explained vividly how the AG interaction worked, and merchandising experts provided fairly elaborate discussions about walking through A and G physically and how it made sense. Given the important phenomenon of effects heredity, A is most likely to be involved. Therefore

Row	A	B	C	D	E	F	G	H	GH DF CE AB	FH DG BE AC	EH CG BF AD	FG DH BC AE	EG CH BD AF	EF BH CD AG	CF DE BG AH	Store 1	Store 2	Average	Range
																Sales change (Units)			
1	-	-	-	-	-	-	-	-	+	+	+	+	+	+	+	-12	-100	-56	88
2	+	-	-	-	+	+	+	-	-	-	-	+	+	+	-	-31	-73	-52	42
3	-	+	-	-	+	+	-	+	-	+	+	-	-	+	-	-51	-74	-62.5	23
4	+	+	-	-	-	-	+	+	+	-	-	-	-	+	+	-43	66	11.5	109
5	-	-	+	-	+	-	+	+	+	-	+	-	+	-	-	-62	-15	-38.5	47
6	+	-	+	-	-	+	-	+	-	+	-	-	+	-	+	30	-38	-4	68
7	-	+	+	-	-	+	+	-	-	-	+	+	-	-	+	-57	-62	-59.5	5
8	+	+	+	-	+	-	-	-	+	+	-	+	-	-	-	178	22	100	156
9	-	-	-	+	-	+	+	+	+	+	-	+	-	-	-	-3	20	8.5	23
10	+	-	-	+	+	-	-	+	-	-	+	+	-	-	+	-16	36	10	52
11	-	+	-	+	+	-	+	-	-	+	-	-	+	-	+	-8	-4	-6	4
12	+	+	-	+	-	+	-	-	+	-	+	-	+	-	-	82	-24	29	106
13	-	-	+	+	+	+	-	-	+	-	-	-	-	+	+	-28	-150	-89	122
14	+	-	+	+	-	-	+	-	-	+	+	-	-	+	-	-25	16	-4.5	41
15	-	+	+	+	-	-	-	+	-	-	-	+	+	+	-	-275	-222	-248.5	53
16	+	+	+	+	+	+	+	+	+	+	+	+	+	+	+	-73	29	-22	102
Effect	77.4	-4.1	-31.1	-20.2	20.4	-2.4	19.8	-25.9	46.3	48.8	9.4	-19.4	-39.1	-70.3	6.7				

The design was generated using E=ABC, F=ABD, G=ACD, H=BCD

FIGURE 5.3
Retailing case: full analysis matrix and aliasing of pair interactions.

the implementation plan was left alone. If the slight ambiguity for now were to be a concern, a 2×2 would resolve it.

So the implementation was A + B – C – D – E + F + G – H–.

5.17 MEASUREMENT PROBLEM FOUND AND FIXED AFTER THE TEST

Hawk-eyed data observers may have noticed with the homogeneity chart that store #6's prior year sales had been recorded in error: #6 in row 3, store 2 (167 had been recorded as 67). The full corrected dataset is in Table 5.2. The re-homogeneity check is only slightly affected and store #29 remains slightly high.

The sales change analysis is unaffected because the prior year had been found in error, not the test data. It was when calculating the change that the error was noticed. The revised pretest biases are: A = –64.0 and AG = 109.6. The earlier pretest reported such biases for A and AG change slightly but the story is the same: neither drove the test results.

5.18 USING SALES CHANGE AS THE TEST'S MEASUREMENT

It's common sense that because stores differ in sales volume, using the change in sales over last year (for the same three calendar months) will level the playing field in that regard. Otherwise the analysis would pick up spurious results tracing back partly to the store volume differences.

Using sales change as the test measurement is important because that is how the industry thinks. The data indicate support for this, in that store volumes differ. An argument might be made for using proportionate sales change but this gives the same findings. The only penalty for using sales change is an increase in noise, due to the subtraction of last year's sales from this year's.

Statisticians might be tempted to run a regression model to check the intervention effects on test sales, with prior year's sales also built into the model. This is ill-advised except to confirm the use of the prior year's data. (It would turn out that the findings are largely unchanged and the

TABLE 5.2

Full Sales Dataset with Store #6 Corrected in Prior Year

Design	Store 1			Store 2			Ranges Between Store Pairs	
Row	Year Ago	Test	Change	Year Ago	Test	Change	R Prior Year	R Test
1	56	44	−12	122	22	−100	66	22
2	49	18	−31	197	124	−73	148	106
3	193	142	−51	167	93	−74	26	49
4	178	135	−43	57	123	66	121	12
5	120	58	−62	72	57	−15	48	1
6	53	83	30	95	57	−38	42	26
7	213	156	−57	106	44	−62	107	112
8	33	211	178	72	94	22	39	117
9	24	21	−3	28	48	20	4	27
10	64	48	−16	46	82	36	18	34
11	73	65	−8	43	39	−4	30	26
12	198	280	82	116	92	−24	82	188
13	76	48	−28	210	60	−150	134	12
14	101	76	−25	65	81	16	36	5
15	859	584	−275	383	161	−222	476	423
16	283	210	−73	114	143	29	169	67

Note: The standard deviation for "year ago" is 63.2 and for the test 39.0 (using adjusted $\bar{R}/1.128$ method) revealing variation tightened in the test.

CALCULATION (METHODS 2 AND 3)

$\bar{R}_{year\ ago} = 1546/16 = 96.63$

Upper limit on R = $3.27\bar{R}$ = 3.27 x 96.63 = 315.96

Removing Row #15:

Adjusted $\bar{R}'$ = (1546 − 476)/15 = 71.33

Adjusted upper limit on R = $3.27\bar{R}'$ = 3.27 x 71.33 = 233.26

No further exclusions.

$\hat{\sigma}_{year\ ago} = \bar{R}'/1.128 = 71.33/1.128 = 63.2$

Repeating for the test (excluding Row #15 then #12) gives:

$\hat{\sigma}_{test} = 39.0$

Noise estimated a year ago (i.e. pre-test) = $\pm t\hat{\sigma}_{year\ ago}\sqrt{(4/N)}$

where t has 15 degrees of freedom after excluding the one range

So, noise = $\pm 2.131(63.2)\sqrt{(4/32)} = \pm 47.6$

Noise estimated from the test = $\pm t\hat{\sigma}_{test}\sqrt{(4/N)}$

where t this time has 14 degrees of freedom since 2 ranges were excluded.

So, noise = $\pm 2.145(39.0)\sqrt{(4/32)} = \pm 29.6$

prior year's sales are indeed significant.) However, it opens up a slippery slope if taken too far.

The care management case, explored with more analytical depth shortly, develops this important caution about mathematical models of randomized orthogonal designs and concludes with useful working rules.

Exercise 15

What would happen if territory sales managers were built into a regression model as a categorical variable? How would this differ from the senior sales managers looking at each intervention and noticing the skills were roughly in balance (all + vs. −)?

5.19 CALCULATING PRECISION AND SAMPLE SIZE BEFORE THE TEST

Before the test runs, its precision or (sensitivity) can be rough-estimated and the sample size confirmed (or increased if inadequate). Using sales data from the year before the test (although in the original work, #6 would have been uncorrected giving slightly differing numbers) and with the randomization into pairs of stores for the test, the test's noise pre-estimates in Table 5.2 are at ±47.6. This means interventions with effects larger in magnitude than 47.6 would be found significant. If this were considered inadequate (with a desire for smaller intervention effects to be visible) then the sample of stores can be increased. The rule here is that each doubling of store quantity will improve the precision (i.e., tighter) by $\sqrt{2}$.

At the time, utilizing 32 stores was judged sufficient. This was a distinct advantage inasmuch as previous tests had traditionally used 75–100 stores, so the cost savings was worth having. The same calculation using the test data gave noise at 29.6. The important insight from this calculation is that the variation between stores subjected to the same interventions dropped in the test. It's not quite significant but often occurs. Therefore, clearly, the pre-estimate of noise and sample size was conservative. This happens more often than not in processes involving people. The pre-test estimates are useful to give a rough idea but will often be pessimistic, such as in this case.

Statisticians also run power calculations to determine in advance the ability of the test to find interventions worth having. Those too will be flawed, in large statistical design work, for the same reason. Again they are a useful rough idea but no more than that.

With hindsight of the test, the power to detect an intervention found just significant would have been 50%: it had had a 50% chance of falling on one side or the other of the noise. For a highly significant intervention, its power would have been higher, approaching 100% above twice the noise.

The twin calculations of power and precision have some use in good statistical design, but not much. They are far more important in other study design types.

A sobering point is that if a small design were run with (say) two or three interventions then the power to detect the almost 20 interventions excluded would be null. The inclusion of sufficient interventions to find some significant is far more germane to successful statistical design than calculations that are unsupported by any theory to carry them into a real-world test, because the variation usually reduces in that test.

5.20 DIAGNOSING UNUSUALLY HIGH OR LOW RESULTS IN A STATISTICAL DESIGN ROW

An important supplement to the primary use of normal plots to find significant effects is to find a single row with an unusually high or low result. There is no other reliable way to do this. If the usual plot looks like an extension ladder leaning against a house, instead of a straight line, an extreme result for one row will be the reason.

This happens because the (say) high result inflates about half the intervention effects where it lands in the + and deflates the other half (by landing in the –). What would have been a roughly bell-shaped curve of insignificant effects centered near zero is split into two parts. Therefore the two straight lines on the normal plot, rather than one, notice this.

This ingenious trick also finds which row is the reason. Figure 2.10 has this split look and by writing down the insignificant intervention effects in order of size, for both sides of the split, gives:

Split 1: EGFB
Split 2: CHD

Especially E and G must have opposite signs than especially C and H in the design (Figure 5.3). Rows 2, 6, 11, and 15 are candidates. Putting F versus D into the mix narrows this down to rows 2 and 15. B won't cooperate but clearly Row 15 has the aberrant result. This row has all the opposites to the optimum, therefore it is low. The result is replicated by two stores quite tightly, therefore it isn't a single aberrant store: it appears real. The two stores are in line with the range of recession drops in the non-test stores, so no harm was done.

Exercise 16

What clue appears in Table 5.2 on the underlying physical cause of the split? (The normal plot usefully confirms the significant effects: they fall outside the split parallels.)

5.21 GUIDANCE ON FRACTIONAL FACTORIAL DESIGNS

The fractional factorial designs are available in the literature [6,8] or can be constructed. They do need expertise to construct correctly. An attractive alternative, where the designs are available in finished form and with less risk of incorrect construction if used with intermediate technical skills, are the multifactorials.

5.22 MULTIFACTORIAL DESIGNS

The multifactorial designs [3,6] were originally developed to test, most efficiently, proximity fuses needed for defense against air attacks over London in the early 1940s. They were designed to find and quantify the significant interventions only. This is a viable strategy because interventions tend to dominate, with interactions (even among pairs) typically found smaller. Later, the facility also to find interactions was added, by working out the aliasing schemes.

The care management case followed that original intent, investigating interventions only. The following goes further into analysis both to prove out the findings already reported and to gain more insight for using the multifactorials. The designs can be found in the referenced material and enough of them appear in this book to start solving any problems.

5.23 CARE MANAGEMENT CASE: MORE ANALYTICAL INSIGHT

The healthcare industry thinks differently from the retail industry. Tests that everyone will implement for competitive advantage therefore also differ, not in the theory but in how they are communicated. Of course, the technical content remains unchanged, as always.

Unlike retail sales managers, care managers do not think in terms of change in admit rates since the same period last year. The clearest correct way to provide results is therefore the raw test and the dry run analyses. No one would know what a change analysis meant, in that industry. In fact, the dry run analysis is not for everyone in care management. It's helpful to senior managers, epidemiologists, healthcare analytics, and so on. It's liable to confuse others because the concept behind the dry run isn't a way people are used to thinking. It's not clear to everyone the distinction between:

1. Analyzing a test before it starts and getting nothing, which seems silly, and
2. Analyzing a test before it starts and getting next to nothing, which is its whole point

Admit rate differences among nurses and hierarchical condition categories (HCC) scores will not affect test findings because they are randomized. The best way to check this is by the shuffling exercise suggested in Chapter 1. Averaging the HCC scores for any random split will find they don't influence the test's findings. Of course they affect the raw numbers, as does every relevant thing in the real world, but not the findings.

If instead of the random shuffling exercise, the average HCC scores were calculated for each intervention's + versus –, then, with the obvious exception of Q, the split differences would fall between –0.1226 and +0.0852, averaging very close to zero. The overall average HCC score for the test is 3.63 with a range of 1.12. These numbers speak for themselves, although the randomization isn't done justice by this clumsy but useful exercise.

As for the retail sales case, it might appear tempting to run some mathematical models and "prove" the findings unaffected by the HCC score or prior admit rate.* If someone tricked regression into doing this (ignoring or adjusting for the problems of theory and practice along the way) then no change in findings would be found; in fact they appear to strengthen.

Exercise 17

How can it be shown that regression to the mean makes no difference to the findings, under randomization? How could regression analysis be tricked into "proving" this? What would its faults be in nonetheless providing the correct answer?

5.24 RANDOMIZATION

The idea of randomization (regarding its usefulness in experimentation) appears to have come to Fisher [1] sometime after an ad hoc 1919 tea-tasting experiment. Gamblers had been using randomization in dice and card games for many years, understanding that it made for fair play. Gamblers understood that it effectively removed any undue influence of the gambling equipment (e.g., dice or cards) leaving the skill of the players to be compared. The same applies in statistical design where influences such as territory salesperson skill, HCC score, or regression to the mean are nullified leaving the intervention effect clear.

Randomization does not eliminate pretest bias in the same way shuffling cards doesn't give everyone the same (or a comparable) hand of cards. This bias issue has been well addressed for statistical design and isn't a problem, provided it is taken into consideration analytically.

To go further into randomization, the randomization distribution [1,4,5,8] is necessary. This isn't important to users in doing good statistical design work, but it is important for technical leaders to gain a fuller

---

* A regression model of the care management data, with $\overline{HCC}$ as a covariate, does not change the findings, except in finding more significance. It's not clear though, what these revised intervention effect estimates <u>mean</u> in the real world. They speculate what might happen in a world devoid of HCC. The pure analysis gives effects expected in the real world, by using the randomization device properly.

appreciation of the randomization device. A short introduction to the randomization distribution is provided in Chapter 8, including its calculation for a couple of the cases. It's folly to try to determine whether a particular randomization did or did not "work." Instead it always will. Also, no one knows what the criterion would be to decide if it had worked. The best that can be done is to poke around such as with the HCC shuffling exercise and see that things are as they should be. The following real examples begin to illustrate the depth behind the randomization device.

5.25 MILK STORY

There's no theory for matched controls in statistical design. This means no one knows what would happen were it to be tried. It is known it wouldn't work and, with luck, we might realize it in time. Gamblers would not hear of a house policy whereby someone arranged the cards in such a way that all hands dealt would be "matched." Also it would not be possible as the real case shown below found out.

Many years ago in England, children were provided a third of a pint of milk every morning as a nutritional supplement during a time of food shortages in the years after WWII. To confirm that the milk helped, a study was devised with a good-sized sample of children getting the milk and an equal sample not. The measurement was weight gain over a defined time.

Rather than randomize, an attempt was made to match the two groups of children. After the test was completed, it was realized that more children from opulent parents were in one of the groups. They tended to have thicker clothes in winter than summer, whereas children from poorer families tended to have similar clothes year round. Couple these flaws with the fact that the study traversed both seasons and any attempt at a pre–post comparison was fatally flawed. The clothes difference was now hopelessly confused with the nutritional effect of the milk when measuring child weight change. A randomized study would have been the only viable design.

It might appear easy to have thought of the opulence issue in advance and included it in the matching algorithm. That's only true with hindsight. Randomization does not require anything be thought of and also anticipates that something(s) will be missed. Randomization effectively "eliminates" all variables from contaminating tests, known or not. Even a little experience with statistical design finds a plethora of things not previously thought of needing real-time redesign or fine-tuning along the way. These appear so obvious with hindsight it looks as if they ought to have been thought of. They won't all be.

The school milk program was eventually discontinued several years after the study by a future prime minister, leading to the media label: "Mrs. Thatcher the Milk Snatcher." The program might well have been discontinued years before had the milk test not been revealed as fatally flawed.

5.26 SOIL STORY

Early use of statistical design in agriculture looked for ways to make plants grow better. So, instead of 20 nurses or 32 retail stores, fields divided into a couple of dozen or so parts were used. A large acreage would be chopped into squares to be used as the test units. These were then randomized to a statistical design to find out what fertilizers and other methods would engender better plants, using some measurable characteristic such as height grown in a certain time or yield (e.g., number of potatoes per plant). Mathematicians at the time felt strongly that randomization would not eliminate a host of nuisance variables such as the gradient of the field at different squares, nutrients already in the soil that varied by square, differing exposure to sunlight, and so on. They argued all those variables should be adjusted in the mathematical model used for analysis. Then and only then, they argued, would the statistical design give correct results and findings. These mistaken views of randomization come from the 1930s and largely remain today. "… [E]xplicit mathematical modeling … was likely to be fraught with hazards; correlations would be complex and unknown and impossible to model adequately. For this reason, at an early stage, [Fisher] had introduced randomization" [5].

Randomization of nurses in the care management design without randomizing members is identical to randomizing by the squares of soil, with both being correct. The care case can sometimes accommodate randomization by patients also, depending on the objective; the soil case cannot. Digging up the field and rescattering soil would be impractical and not real world.

Randomization differed from its detractors in devising a method usable in the real world, especially given that the proposed mathematical adjustments would be impossible but its flaws not noticed. The banking industry's near collapse of the world economy by 2008 had in part been caused by misapplied mathematical models. By the time the models were put into practice, few could understand they were wrong. Ironically, those who could understand also did not realize in time they were wrong. So, Fisher rightly foresaw that real people solving real problems would need to be able to use a foolproof method.

Mathematical models are useful but always a little (or a lot) off. A useful exercise is to use them on any of the cases in this book and try to "break" the results and findings without violating any theoretical requirements of the models.

In one retail case it was asked that sales trends which stores were already on be backed out to make sure the pure intervention effects were calculated. Pragmatically it made sense to analyze both ways and show no substantial difference, so that later cases could use the pure randomization device with everyone seeing it worked on their data. Therefore it's useful to run the mathematical models now that the software is readily available. They'll usually re-prove the purer findings of the randomized design, with about half of the effects appearing a little larger and the other half a little smaller. In a few cases the models will be off.

5.27 GEOMETRIC VERSUS NONGEOMETRIC DESIGNS

The multifactorial designs accommodate almost all practical problem-solving needs, especially intervention quantities up to the few dozen that are rarely exceeded by practical needs. For cases where more than 95 interventions need to be tested (which is very rare) the fractional factorials have no upper limit, except again the practical constraint that too many will become unmanageable, depending on case specifics as to how many, usually two to four dozen.

The multifactorials divide into two types: geometric (having their number of rows as a power of 2: the 8, 16, 32, and 64-run designs) and nongeometric (which occur in row quantities as multiples of 4: these start with the 12, 20, 24, 28, and 36-run designs). There is an advantage in using some of the nongeometric designs in view of their aliasing schemes. Interventions tend to dominate so any multifactorial would apply, however, the nongeometric designs have complex aliasing schemes that can bring an advantage. This advantage was the reason for choosing the 20-row design in the care management case.

When using the multifactorials as originally intended (to estimate interventions effects only), the preference is for nongeometric designs for the reasons next explained. When using the multifactorials to also give interaction effects, akin to the retail case design, a simple augmentation to the basic multifactorial designs is needed, using a technique called foldover (see Section 5.29).

5.28 ALIASING SCHEME FOR THE CARE MANAGEMENT DESIGN

The aliasing schemes for the nongeometric designs listed above (12, 20, 24, 28, 36 runs) are complex. Given that interaction effects are at the outset likely to be smaller than intervention effects (and therefore the interventions dominate), the complex aliasing scheme amplifies that dominance. Even a significant interaction will be spread among more than one intervention column, making its influence more slight. The nongeometric designs can seem disadvantageous because it appears they will not estimate interactions as well, if at all. Again though, the primary purpose is the interventions themselves, protected reasonably against being contaminated especially by pair interactions.

The retail case shows that the dual objective (of interventions and their pair interactions) is straightforward. That's not always practical given the business problem and, although statistical designs reveal a vast amount, quickly and thoroughly, it's not realistic or important to find out everything at once.

If it were supposed that testing a few interventions would be safer, the improvement would be put at risk, as this chapter's opening mathematical proof noted. So it's really not safer but far riskier. It would just satisfy a less important preference for easier mathematics behind the scenes.

5.29 AUGMENTING MULTIFACTORIALS TO ALSO ESTIMATE PAIR INTERACTIONS

The multifactorial designs are easily augmented to behave as the retail case design did in estimating pair interactions quite well. The geometric designs,

Run	A	B	C	D	E	F	G	H	I	J	K	L	M	N	O
1	+	-	-	-	+	-	-	+	+	-	+	-	+	+	+
2	+	+	-	-	-	+	-	-	+	+	-	+	-	+	+
3	+	+	+	-	-	-	+	-	-	+	+	-	+	-	+
4	+	+	+	+	-	-	-	+	-	-	+	+	-	+	-
5	-	+	+	+	+	-	-	-	+	-	-	+	+	-	+
6	+	-	+	+	+	+	-	-	-	+	-	-	+	+	-
7	-	+	-	+	+	+	+	-	-	-	+	-	-	+	+
8	+	-	+	-	+	+	+	+	-	-	-	+	-	-	+
9	+	+	-	+	-	+	+	+	+	-	-	-	+	-	-
10	-	+	+	-	+	-	+	+	+	+	-	-	-	+	-
11	-	-	+	+	-	+	-	+	+	+	+	-	-	-	+
12	+	-	-	+	+	-	+	-	+	+	+	+	-	-	-
13	-	+	-	-	+	+	-	+	-	+	+	+	+	-	-
14	-	-	+	-	-	+	+	-	+	-	+	+	+	+	-
15	-	-	-	+	-	-	+	+	-	+	-	+	+	+	+
16	-	-	-	-	-	-	-	-	-	-	-	-	-	-	-
17	-	+	+	+	-	+	+	-	-	+	-	+	-	-	-
18	-	-	+	+	+	-	+	+	-	-	+	-	+	-	-
19	-	-	-	+	+	+	-	+	+	-	-	+	-	+	-
20	-	-	-	-	+	+	+	-	+	+	-	-	+	-	+
21	+	-	-	-	-	+	+	+	-	+	+	-	-	+	-
22	-	+	-	-	-	-	+	+	+	-	+	+	-	-	+
23	+	-	+	-	-	-	-	+	+	+	-	+	+	-	-
24	-	+	-	+	-	-	-	-	+	+	+	-	+	+	-
25	-	-	+	-	+	-	-	-	-	+	+	+	-	+	+
26	+	-	-	+	-	+	-	-	-	-	+	+	+	-	+
27	+	+	-	-	+	-	+	-	-	-	-	+	+	+	-
28	-	+	+	-	-	+	-	+	-	-	-	-	+	+	+
29	+	-	+	+	-	-	+	-	+	-	-	-	-	+	+
30	+	+	-	+	+	-	-	+	-	+	-	-	-	-	+
31	+	+	+	-	+	+	-	-	+	-	+	-	-	-	-
32	+	+	+	+	+	+	+	+	+	+	+	+	+	+	+

FIGURE 5.4
Multifactorial design for 15 interventions, with foldover.

having clean aliasing schemes, are easier for this purpose. *Foldover* [8] simply doubles up the design (e.g., from 16 to 32 runs) but in the additional runs changes the signs: – to + and + to –. Figure 5.4 shows the foldover design for 15 interventions in 32 rows. By analyzing the full design with foldover (e.g., the 32 runs in Figure 5.4) the intervention effects are without any aliasing with pair interactions. The pair interactions can then also be estimated; a real case illustrates in Chapter 7.

5.30 TESTING STRATEGY

To conclude with an important point touched on in the cases a few times already: a good strategy for problem-solving, if to be measured by something improving suddenly then sustaining, has been found as illustrated in the cases. If the rules are otherwise then a refining test or perhaps a randomized control trial (RCT) at implementation may satisfy. In running a large statistical design followed by immediate implementation, statistical control becomes completely important, as a real-time feedback-control device as well as carrying the administrative task of showing and quantifying the improvement and its amount, in comparison to the earlier prediction.

This strategy recognizes that, pragmatically, it is competitive to learn a lot, fast and well, but not everything at once. It also recognizes that the optimum solutions keep shifting over time and no statistical design (or set of them) will know about future shifts. Statistical control integrated with statistical design provides a practical way to adapt the design's findings over the longer term. Therefore the idea is to optimize the solution today and adapt it tomorrow and then continuously into next year. This also underlines why statistical control has to be quick and easy as people will only have a few minutes a day to use it.

Analytically, if change since pretest is to be used in some form, a graph of test versus pre will confirm. A regression finding the pre significant is quicker and also useful at analysis to reassure that no other variable influenced the findings. Regression can be taken too far, becoming fraught with complexities the randomization is designed to overcome. A useful rule of thumb in analysis is that if it couldn't be done (or reproduced) by hand, it probably shouldn't be.

Realistically, time management requires statistical software to be used. It should stay close, however, to the simple construct that orthogonal design just compares test versus counterfactual and the randomization, if left untampered with analytically, gives a direct window into the real world.

5.31 UNIQUENESS AND STUMBLING AROUND

Every problem is unique. Therefore cases are not entirely helpful as each is more different than similar. The unsuspecting scientist, succumbing to a request for case examples to explain how a potential new case might work, will have this pointed out for him if he's not careful! It is helpful though for most people to see something that's along the same lines as what they need to solve. As differences are pointed out, it's most productive to throw out the canned case and start working on theirs, in the moment.

Because each problem is unique, it turns out that experienced scientists ought to be able to show up in an industry or with a problem type they've never seen before and, with no preparation, operate in formal meetings and initial research, just the same as if it's one they've already cracked many times. There might be a little fumbling with language but nothing substantial should impede the work at full power and speed.

Because new ground is always being plowed with each case, there's a certain amount of stumbling around to find the right solution approach. Improvement and innovation are not perfectly packaged and templated like clockwork. Each is to an extent improvised, using rigorous rules. With the care management and retail cases, there may be the appearance of perfect choreography and execution. It wasn't that way. Each case is new with its own hidden foibles that are stumbled into and dealt with from a vast arsenal of tactics built through experience and good theory/mathematics.

The same cases illustrate that the stumbling around is most uncertain at the implementation phase. It's impossible to predict what will happen and need action to realize the full improvement predicted by the design's analysis. Being ready, with an eye on the statistical control charts, is invariably viable.

5.32 WHERE DID STATISTICAL DESIGN ORIGINATE?

On March 20, 1923, a paper was published with the analysis of a field trial investigating potato growing. Noticing that the trial's layout had been poor, an industrial statistician in the brewing industry, W. S. Gosset, remarked to R.A. Fisher*, a scientist at Rothamsted Research (formerly Rothamsted Experimental Station): "I think you should help them with that." The text [1] that introduced statistical design described early in its pages an impromptu 1919 tea-tasting experiment. It appears that both events, four years apart, sparked statistical design.

---

* Fisher's more advanced contributions remain ahead of their time. The discovery of orthogonal design, the use of the randomization device, and making the early designs work in practice, were extraordinary accomplishments.

Important developments then occurred with the fractional factorials a few years later and the multifactorials in the 1940s with foldover added a few years after that. The more modern realization that replication was less important than first practiced began to develop by the 1960s, and the proof that larger designs brought more improvement was published in 1966. Effects sparsity and heredity followed: the empirical realization that among many interventions tested, only a few would be found significant, and those few would be more likely to interact.

The design used in the care management case is sometimes called *saturated* inasmuch as it fills it up with the maximum number of interventions (19 in 20 runs). Designs that test more interventions than the number of runs (called *supersaturated designs* [6]) were made possible by effects sparsity but require considerable mathematical expertise and carry the disadvantage of losing their intuitive connection to nontechnical users who need to know roughly how the design works before they'll agree to run or implement it.

Over the years, more theory has been added, especially to enjoy advantage at the margin of one design choice over another. These advantages are often in the aliasing schemes and in setting up the design for most effective analysis. In practice, the fundamentals deployed in an organization will bring more competitiveness than perfect mathematical structures at the margin. In practice, there is often just a few minutes to assemble a design as the last intervention rolls off the creative pipeline and with the copier ready to roll on a neat set of procedures for everyone to kick off by their early morning meeting. It's also been found that a flood taking out a basement test unit, or a severe weather system challenging a utility line test, or people in a counterfactual taking a liking to what the test people got for a particular intervention, dominate in the real world.

Exercise 18

If more people in the counterfactual than those who were supposed to tested an intervention, could the resulting mess be cleaned up? How could this be done with no further analytics but with mental arithmetic in a second or two? Are the test and its results still valid? Should the test have been better managed? Or was it?

How could this have happened, when clear education in the test procedures was provided and agreed to by all?

6

Statistical Design and Control: A Dozen Large-Scale Case Studies

6.1 SELECTION OF CASES

Statistical design and control both require proof from businesses improving suddenly then sustaining. More information has been provided by now about why there is no mathematics to predict whether a particular completed statistical design will, or will not, improve a process, until it does. Statistical control recognizes there is also no mathematics to explain (or predict) why a process improves (or by how much) when it is stabilized. The method is therefore designed to exploit the real world, in real-time, as it unfolds. There is no theory to calculate sample sizes except roughly, in statistical design, because the variation often tightens. Comparisons of variation before and during testing are given in several cases.

Real cases are fraught with imperfections, adaptations, and course corrections improvised in the moment. Those real-world cases are the intent here, without compromising the mathematics or rules of objective presentation of data. All but two of the dozen cases in this book were selected from a two-year period ending in early 2014. The exceptions come from 1980 and 1998 (the backpacking case). Two of the dozen cases appeared in the opening chapters, another in Chapter 5, and the final one follows in Chapter 7.

Most of the cases are simplified from the original work, for publication, but without altering the technical content. The dozen cases were also selected to represent a diverse range of problem types and industries. Together they point the way for any business problem needing solution. The standard format is self-explanatory using the main methods and displays already introduced. Narrative is restricted to remarks at the end of each case, bringing out useful features that can be reused, or problems that can easily be avoided next time. Some cases have no implementation data. Two were research projects and one is in progress at the time of going to print (the orthogonal conjoint).

Significances (shaded in results) can be checked by Method 3, normal plot or regression. In the long run over many cases, about half of interventions helping and half hurting have been found. Not all are significant. In the cases here it will be found that 21% of the total interventions significantly help and 7% significantly hurt (excluding interactions). This 3:1 ratio varies by project and by experimenter.

Exercise 19

How would large statistical design look on:

a. Software design to release with fewest bugs.
b. Website/SEO (search engine optimization) optimization to maximize sales.
c. Reducing power outages due to trees and squirrels.
d. Redesigning an artistic layout to provide the highest read rate as measured by follow-up consumer actions.
e. Reducing critical events (e.g., infrastructure power outages).
f. Newsprint or online media layout and content to maximize sales and advertiser benefit.
g. Reducing regulatory complaints where any of five call centers and direct channels may have caused or festered the eventual complaint a few weeks later. By that time the roots of the complaints are a few weeks old.
h. First-time call resolution in a service center.
i. Reduce repeat calls to a call center.
j. Reduce install time influenced by field technician crews and central offices (CO) with the circuitry. Assume three crews per dedicated CO.
k. Cross-channel optimization in retail with 100 stores and online shopping to maximize sales without cannibalizing.
l. Redesign an online payment system to minimize expensive questions called in to a center.

1. Call Center Productivity, Consumer Sales to Highly Qualified Leads.

Objective: increase sales productivity without reducing % leads sold.

Measurements: calls per week
% of leads sold

Test units: sales agents

Measurement error:

Service calls excluded to measure pure sales productivity. 13.5% of variation in % leads sold, stable.

Homogeneity:

Agent (random order)

[Individuals chart; c = chart oversensitive given call volumes]

[x̄, R charts] Agent Pair (as randomized for test)

Interventions:

A: Simple script, 1/4 length
B: Assumptive ID validation
C: Drop lead history
D: Explain advantage
E: No co-branding
F: Agent feedback daily on
    productivity
G: Stress value over "new"
H: Only mention brand once
I: Endorse with "I recommend"
J: Drop survey
K: Use plain language not
    industry jargon
L: "Needs" script not service
    advantage
M: State delivery time
N: Ask early
O: Drop maintenance
    advantage
P: Discuss user (not service)
    advantages

Design and Results:
Design: Multifactorial 20-runs, fully replicated, 40 full-time agents over 4 weeks

Row	A	B	C	D	E	F	G	H	I	J	K	L	M	N	O	P	Q	R	S	Agent 1 Calls Pre	Agent 1 Calls Test	Agent 2 Calls Pre	Agent 2 Calls Test	Average Calls Pre	Average Calls Test	Range Calls Pre	Range Calls Test
1	+	−	−	+	+	+	+	+	−	+	+	+	−	−	−	−	+	+	−	1620	2195	2982	2457	2301	2326	1362	262
2	+	−	−	+	+	+	+	+	+	+	+	−	−	−	−	+	+	+	+	2132	2060	1259	2117	1695.5	2088.5	873	57
3	−	−	+	+	+	−	+	−	−	+	+	+	−	+	+	−	+	+	−	2200	2343	1174	1225	1687	1784	1026	1118
4	−	+	+	+	−	+	+	−	+	+	−	−	+	+	−	+	−	+	−	1126	1048	1181	626	1153.5	837	55	422
5	+	+	+	+	+	+	−	+	+	−	−	+	+	−	+	−	+	−	+	1858	1778	1329	841	1593.5	1309.5	529	937
6	+	+	+	−	+	+	+	−	+	+	+	+	+	+	−	+	−	+	−	568	950	1237	1409	902.5	1179.5	669	459
7	+	+	+	+	−	−	−	+	+	+	+	+	+	−	+	−	−	+	−	1517	1561	936	850	1226.5	1205.5	581	711
8	+	+	−	+	+	+	+	+	−	+	+	−	−	+	+	+	+	−	+	304	561	2005	1558	1154.5	1059.5	1701	997
9	+	+	+	+	+	−	+	+	+	−	+	+	−	+	+	−	+	+	+	937	1065	2003	1234	1470	1149.5	1066	169
10	+	−	+	−	−	+	−	+	+	+	+	+	+	+	+	+	+	+	−	701	1095	2011	1612	1356	1353.5	1310	517
11	+	+	−	+	+	+	+	+	+	+	−	+	−	−	+	+	−	−	−	1364	1204	2296	2023	1830	1613.5	932	819
12	+	−	+	+	−	+	+	+	+	+	+	−	+	+	+	+	+	+	+	1175	2452	3216	3036	2195.5	2744	2041	584
13	+	+	+	−	+	+	−	+	−	+	+	−	+	+	+	+	+	+	−	1509	1278	249	481	879	879.5	1260	797
14	−	−	−	−	+	+	+	+	+	+	−	+	+	+	−	−	−	+	−	2040	1410	1375	921	1707.5	1165.5	665	489
15	−	+	+	+	+	−	−	−	+	+	+	−	−	+	−	+	+	−	−	2042	2455	1602	1070	1822	1762.5	440	1385
16	+	+	+	+	+	+	−	−	+	−	−	+	+	+	+	−	+	+	+	863	759	1565	1089	1214	924	702	330
17	−	+	−	−	+	+	+	−	+	+	+	−	+	+	+	−	+	−	+	1884	2422	1002	2020	1443	2221	882	402
18	+	−	−	+	−	+	−	+	+	+	−	−	+	−	−	+	+	+	+	941	888	835	825	888	780.5	106	89
19	+	−	+	−	+	−	+	+	−	−	+	+	+	+	−	+	−	+	−	1593	1616	2323	2257	1958	1936.5	730	641
20	−	−	−	−	+	−	−	−	+	+	−	−	+	+	+	−	+	−	+	575	631	1142	1029	858.5	830	567	398

R̄=874.9   579.2

Noise 511.6   338.7

Dry Run: Calls	A 17.7	B 84.9	C −187.8	D 3.8	E −106.1	F 380.9	G 357.2	H 195.1	I −83.1	J 267.5	K 147.9	L −180.3	M 91.6	N −165.2	O −34.6	P 196.8	Q 117.5	R 308.4	S −195.4
Test: Calls	339	25.5	−330	−54.25	−22.05	477.1	371.8	159	67.9	358	151.1	−369.8	166.5	−411.3	201.1	147	−39.2	197.25	−179.7

Change in σ̂ (among agents, per month)

	Pre	Test
σ̂$_{calls}$	776	513

Implementation: ABFIJ - Increases calls and/or reduces work.

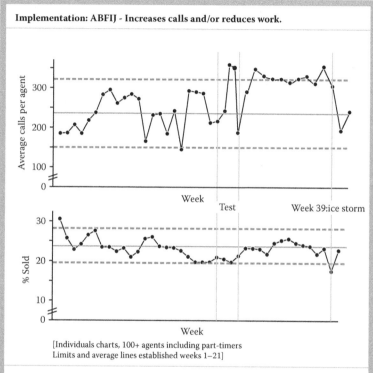

[Individuals charts, 100+ agents including part-timers
Limits and average lines established weeks 1–21]

Narrative:

The product enjoyed the highest close rate among several lines; productivity increased units sold.

The original design included 3 more interventions, dropped as the test started, to make agent task easier.

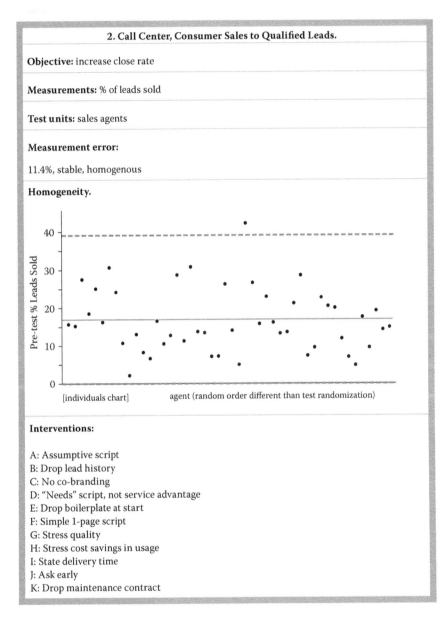

2. Call Center, Consumer Sales to Qualified Leads.

Objective: increase close rate

Measurements: % of leads sold

Test units: sales agents

Measurement error:

11.4%, stable, homogenous

Homogeneity.

Pre-test % Leads Sold

[individuals chart]   agent (random order different than test randomization)

Interventions:

A: Assumptive script
B: Drop lead history
C: No co-branding
D: "Needs" script, not service advantage
E: Drop boilerplate at start
F: Simple 1-page script
G: Stress quality
H: Stress cost savings in usage
I: State delivery time
J: Ask early
K: Drop maintenance contract

Design and Results:

Design: Multifactorial 12-run with foldover, fully replicated, 48 full-time agents over 4 weeks

Row	A	B	C	D	E	F	G	H	I	J	K	Agent 1 Calls Pre	Agent 1 Calls Test	Agent 1 %leads Sold Pre	Agent 1 %leads Sold Test	Agent 2 Calls Pre	Agent 2 Calls Test	Agent 2 %leads Sold Pre	Agent 2 %leads Sold Test	Average Calls Pre	Average Calls Test	Average %leads Sold Pre	Average %leads Sold Test	Range Calls Pre	Range Calls Test	Range %leads Sold Pre	Range %leads Sold Test
1	−	+	−	−	−	+	+	+	−	+	+	394	188	15.7	19.7	43	14	14.0	14.3	218.5	101.0	14.8	17.0	351.0	174.0	1.8	5.4
2	+	−	−	−	+	+	+	−	+	−	+	98	307	15.3	14.7	40	184	5.0	1.6	69.0	245.5	10.2	8.1	58.0	123.0	10.3	13.0
3	−	−	−	+	+	+	−	+	+	+	−	102	113	27.5	23.0	81	54	43.2	18.5	91.5	83.5	35.3	20.8	21.0	59.0	15.8	4.5
4	−	−	+	+	+	−	+	+	+	−	−	545	12	18.5	25.0	30	47	26.7	46.8	287.5	29.5	22.6	35.9	515.0	35.0	8.1	21.8
5	−	+	+	+	−	+	+	−	−	−	+	195	327	25.1	9.5	19	38	15.8	21.1	107.0	182.5	20.5	15.3	176.0	289.0	9.3	11.6
6	+	+	+	−	+	+	−	−	−	+	−	105	105	16.2	19.0	314	150	23.9	27.3	209.5	127.5	20.0	23.2	209.0	45.0	7.7	8.3
7	+	+	−	+	+	−	−	−	+	−	+	166	151	30.7	27.2	310	297	16.1	14.5	238.0	224.0	23.4	20.8	144.0	146.0	14.6	12.7
8	+	−	+	+	−	−	−	+	−	+	+	29	43	24.1	27.9	514	446	13.2	14.3	271.5	244.5	18.7	21.1	485.0	403.0	10.9	13.6
9	−	+	+	−	−	−	+	−	+	+	−	169	313	10.7	10.2	398	72	13.6	29.2	283.5	192.5	12.1	19.7	229.0	241.0	2.9	18.9
10	+	+	−	−	−	+	−	+	+	−	+	91	47	2.2	12.8	19	88	21.1	6.8	55.0	67.5	11.6	9.8	72.0	41.0	18.9	5.9
11	+	−	−	−	+	−	+	+	−	+	+	200	369	13.0	13.6	21	5	28.6	20.0	110.5	187.0	20.8	16.8	179.0	364.0	15.6	6.4
12	−	−	−	+	−	+	+	−	+	+	+	24	99	8.3	11.1	544	76	7.4	11.8	284.0	87.5	7.8	11.5	520.0	23.0	1.0	0.7
13	+	−	+	−	+	+	−	+	−	−	−	240	195	6.7	8.2	537	342	9.5	7.9	388.5	268.5	8.1	8.0	297.0	147.0	2.8	0.3
14	−	+	−	+	+	−	+	−	−	+	−	176	58	16.5	24.1	97	44	22.7	36.4	136.5	51.0	19.6	30.3	79.0	14.0	6.2	12.2
15	+	−	+	+	−	+	−	−	+	+	−	86	48	10.5	27.1	42	119	21.4	17.6	64.0	83.5	15.9	22.4	44.0	71.0	11.0	9.4
16	−	+	+	−	+	−	−	+	+	−	−	173	280	12.7	9.3	140	55	20.0	16.4	156.5	167.5	16.4	12.8	33.0	225.0	7.3	7.1
17	+	+	−	+	−	−	+	+	−	−	+	174	14	28.7	28.6	438	207	11.9	19.3	306.0	110.5	20.3	23.9	264.0	193.0	16.9	9.2
18	−	−	+	−	−	+	+	+	+	+	+	248	29	11.3	10.3	413	228	7.0	7.5	330.5	128.5	9.2	8.9	165.0	199.0	4.3	2.9
19	+	+	+	+	+	+	+	+	+	+	+	39	87	30.8	18.4	673	280	4.9	5.4	356.0	183.5	17.8	11.9	634.0	193.0	25.9	13.0
20	−	+	+	−	−	+	−	−	−	−	−	97	3	23.7	66.7	63	23	17.5	56.5	80.0	13.0	20.6	61.6	34.0	20.0	6.3	10.1
21	−	−	+	−	+	−	−	−	−	+	+	64	20	23.4	30.0	42	63	9.5	20.6	53.0	41.5	16.5	25.3	22.0	43.0	13.9	9.4
22	−	+	−	−	+	+	+	+	−	−	−	127	378	7.1	4.2	115	129	19.1	38.8	121.0	253.5	13.1	21.5	12.0	249.0	12.0	34.5
23	+	−	−	+	−	+	+	−	−	−	−	316	138	7.3	15.9	135	121	14.1	10.7	225.5	129.5	10.7	13.3	181.0	17.0	6.8	5.2
24	+	+	−	−	−	−	−	−	+	+	−	80	37	26.3	21.6	148	126	14.9	23.8	114.0	81.5	20.6	22.7	68.0	89.0	11.4	2.2
R																					199.7	141.8					10.1	8.9

Noise = 105.5 5.32 (Calls)   74.9 4.69 (Sold)

	A	B	C	D	E	F	G	H	I	J	K
Dry Run:											
Calls	73.3	7.0	−25.8	18.4	−0.3	70.6	0.9	−19.4	−36.9	1.0	90.0
% Sold	0.25	−3.94	−1.38	−0.31	0.43	−0.77	−2.06	−1.34	3.05	−5.39	−3.68
Test:											
Calls	59.4	−4.9	16.3	−56.0	13.7	34.4	−22.9	−70.2	27.5	−38.9	
% Sold	−4.20	−6.82	4.67	0.30	3.36	3.03	−8.96	5.40	3.91	−5.76	−2.61

Change in σ̂ (among agents, per month)

	Pre	Test
σ̂ calls	177	126
σ̂ sales %	8.9	7.9

Implementation: C, H - increases % sold.

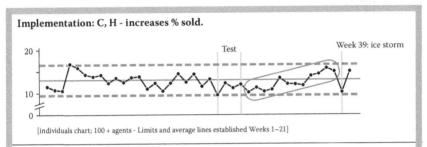

[individuals chart; 100 + agents - Limits and average lines established Weeks 1–21]

Narrative:

A slight non-homogeneity was included.

The product being sold had been the most problematic for agents. Accordingly the inter-ventions were cut to 11 as the test started. No attempt was made to pick the strongest. The trend in the 14 weeks since the test is significant even including the ice storm setback.
As this and the previous case implemented, a new statistical design was prepared to extend both improvements to all products. Preparing baseline data found the following:

All Products

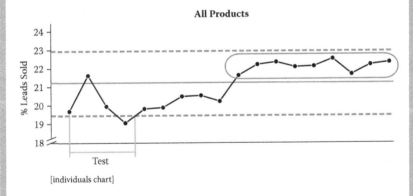

[individuals chart]

By remote monitoring of calls and talking to supervision, it turned out agents had been using helpful interventions from the pair of tests on everything, for a while. The extra test was cancelled.
The final chart was prepared with hindsight. It could be argued a chart split at the sudden increase (confirmed by the run of 9 above the average, later) would be more appropriate, and show the improvement more clearly. This chart is reproduced since it was the one done. This pair of call center cases also reveal why percentage measurements need care in analysis, since the interventions may increase productivity. So the denominator can't be used blindly.

3. Direct Marketing
Objective: response rate
Measurements: proportion of current users upgrading service
Test unit: 500 customers randomly selected ($\times$ 28 test units)
Measurement error: independent data pulls at 30 and 45 days
Homogeneity: not needed where individual customers are randomized. This ensures homogeneity, similar to a randomized control (clinical) trial.
Interventions: A: Co-branded B: "Sell hole vs. drill" picture C: Outer envelope in color D: 800 # on front E: Price reduction F: Blue vs. gray dominant color G: Call to Action - "limited time offer" language H: Placement of Call to Action I: Pictures of users included J: Personalized K: Return mailer included L: Website featured M: Phone number prominent N: Family picture included O: Guide included P: Local agent card included Q: Specific piece on gap closure R: Paper color S: Calendar magnet included T: QR scan U: Find us on Facebook V: Mailed priority vs. standard W: Breakdown of costs without insurance included X: Product bought previously Y: Letter from local agent included Z: Focus on one provider vs. many AA: Yellow sticky tab graphic included

Design and Results:
Design: multifactorial 28-runs

Row	A	B	C	D	E	F	G	H	I	J	K	L	M	N	O	P	Q	R	S	T	U	V	W	X	Y	Z	AA	Response
1																												0.0089
2																												0.0036
3																												0.0533
4																												0.0044
5																												0.0622
6																												0.0133
7																												0.0133
8																												0.0036
9																												0.0578
10																												0.0178
11																												0.0089
12																												0.0089
13																												0.0053
14																												0.0578
15																												0.0116
16																												0.0169
17																												0.0116
18																												0.0444
19																												0.0053
20																												0.0107
21																												0.0142
22																												0.0667
23																												0.0009
24																												0.0533
25																												0.0027
26																												0.0116
27																												0.0098
28																												
Effect	0.0023	0.0081	-0.0045	0.0062	-0.0017	0.0017	0.0035	0.0093	-0.0021	0.0017	-0.0017	-0.0060	-0.0015	0.0015	-0.0030	-0.0012	0.0045	-0.0012	-0.0002	-0.0043	-0.0002	-0.0012	-0.0269	0.0037	0.0260	0.0058	0.0030	0.0006

Change in δ(among campaigns): see test phase on implementation chart

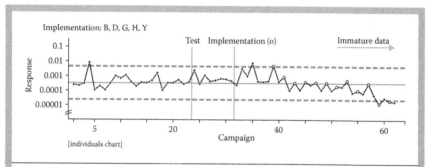

Narrative:

The direct industry feels strongly each campaign is different whereas statistical control finds them mostly stable. Scaling in increments of .1, .01, .001, etc. makes this easier to see. A randomized control trial (splitting a campaign into implementation vs. control, or "challenger" vs. current "champion" as it is also known) is the only proof accepted by tradition. Yet the split is impractical except in testing. Also the slight improvements that bring stronger market response are not usually going to break control limits. Alternating implementation (o) vs. control (•) clears this up, cheaply and clearly. The (o) are usually (not always) higher than their neighbors (•).

The implementation lags a few weeks. The fading response as an artifact of immature data is seen near the end (after about campaign #40 and more sharply by #60) as the data was cut too recently.

The test saw a higher response, using a random sample of a larger campaign.

4. Electronic Component Design.

Objective: maximize failure voltage of a component in a circuit design, while minimizing its dispersion (variation)

Measurements: log failure voltage in a destructive test
range among two measurements

Test units: microchips subjected to a destructive test

Measurement error:

N/A in a destructive test. Lab standards were tightened to minimize. The risk was that if measurement error were excessive no significant results would be found.

Homogeneity:

N/A in production processes. Instead stability over consecutive components was controlled by using components from the same lot. Stability is then checked by the limit on the ranges during the test. One instability found (see design results).

Interventions:
A: Component design: 1 vs. 2
B: Test equipment design: 1 vs. 2
C: Circuit design: 1 vs. 2

Design and Results:
Design: Full factorial.

Row	A	B	C	AB	AC	BC	ABC	y1	y2	y-bar	R
1	-	-	-	+	+	+	-	2.301	2.301	2.301	0
2	+	-	-	-	-	+	+	3.255	2.845	3.050	0.410
3	-	+	-	-	+	-	+	3.255	2.903	3.079	0.352
4	+	+	-	+	-	-	-	3.342	3.146	3.244	0.196
5	-	-	+	+	-	-	+	2.602	2.602	2.602	0
6	+	-	+	-	+	-	-	2.699	2.845	2.772	0.146
7	-	+	+	-	-	+	-	3.477	3.342	3.410	0.135
8	+	+	+	+	+	+	+	3.663	3.663	3.663	0
1	-	-	-	+	+	+	-	2.301	2.602	2.452	0.301
2	+	-	-	-	-	+	+	2.699	3.255	2.977	0.556
3	-	+	-	-	+	-	+	2.845	3.255	3.050	0.410
4	+	+	-	+	-	-	-	3.342	3.342	3.342	0
5	-	-	+	+	-	-	+	2.602	2.477	2.540	0.125
6	+	-	+	-	+	-	-	3.531	2.477	3.004	1.054
7	-	+	+	-	-	+	-	3.342	3.146	3.244	0.196
8	+	+	+	+	+	+	+	3.663	3.342	3.503	0.321
1	-	-	-	+	+	+	-	2.778	2.699	2.739	0.079
2	+	-	-	-	-	+	+	3.342	3.415	3.379	0.073
3	-	+	-	-	+	-	+	3.146	3.146	3.146	0
4	+	+	-	+	-	-	-	3.477	3.342	3.410	0.135
5	-	-	+	+	-	-	+	2.602	2.602	2.602	0
6	+	-	+	-	+	-	-	2.602	2.903	2.753	0.301
7	-	+	+	-	-	+	-	2.954	2.954	2.954	0
8	+	+	+	+	+	+	+	3.477	3.663	3.570	0.186
Effect:	0.379	0.537	0.037	-0.071	-0.060	0.141	0.128	Noise = ± 0.090		$\bar{R}'$ =	0.171
Dispersion Effect:	0.148	-0.093	-0.004	-0.191	0.110	-0.039	-0.009				

Implementation:

Component design 2 and test equipment 2 (A+B+) which also gives the AB interaction, reducing dispersion. Because of the BC interaction, circuit design 2 (C+) was also used. More strength came from the ABC intervention.

Narrative:

This maximized the strength and consistency of the components, to shocks from electrostatic discharge in spacecraft, where repair was difficult.
The logarithms of voltages are shown, used because raw data were highly skewed.
The dispersion analysis just used R as the measurement.
The one instability (the range of 1.054) did not change findings or significance (except to weaken the AB dispersion finding).

---

5. Readmits Reduction.

Objective: reduce rehospitalizations for recently discharged patients

Measurement: admits per thousand members per year

Test units: case managers

Measurement error:

18%, stable and homogeneous

Homogeneity:

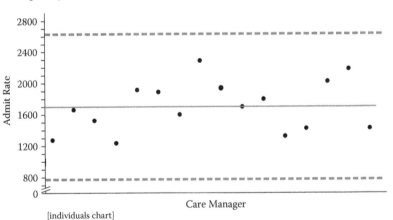

[individuals chart]

Interventions:

A: Weekly meeting; case reviews
B: Care manager feedback
C: Falls prevention
D: Transition report
E: 1st session within 7-10 days
F: Post-discharge education
G: Home visit

Design and Results:

Design: Multifactorial 8-runs with foldover

Row	A	B	C	D	E	F	G	pre	test
1	+	+	+	-	+	-	-	1276	1651
2	-	+	+	+	-	+	-	1661	1504
3	-	-	+	+	-	-	+	1521	1291
4	+	-	-	+	+	-	-	1234	1409
5	-	+	-	+	+	+	+	1918	976
6	+	-	+	-	+	+	+	1896	1269
7	+	+	-	+	-	+	+	1607	1581
8	-	-	-	-	-	-	-	2295	1953
9	-	-	-	+	-	+	+	1948	1189
10	+	+	-	-	+	-	+	1711	1331
11	+	+	-	-	-	+	-	1800	1207
12	-	-	+	-	-	-	+	1320	1810
13	+	-	+	+	-	-	-	1429	1855
14	-	+	-	+	+	-	-	2021	1552
15	-	-	+	-	+	+	-	2187	1252
16	+	+	+	+	+	+	+	1426	870
Test	-44.25	-49.75	38.00	-24.75	-254.50	-418.50	-258.25		
Dry Run	-311.50	-149.00	-227.25	-194.50	-82.75	111.25	-69.50		

Implementation: E, F, G

Narrative:

The post-discharge education (F) had been postponed at the "last minute". The finding that it "helped" called the whole test into question, until discussion and records showed the session had been scheduled, announced, and the manual distributed for pre-reading. Further checks suggested the manual had been read as required and found useful.

No significant interactions were found among E, F, and G as most likely.

It had previously been difficult to analyze whether care management helped. That it could be improved meant it helped in the first place, now more so.

There were no significant interactions.

---

6. Retention Letter Design (Orthogonal Conjoint).

Objective: increase retention in insurance industry

Measurements: % retention
                survey

Test units: randomly allocated samples of existing clients

Interventions:

A: White paper on risks included
B: C-level signature
C: Snazzy package
D: Address without first name
E: Include recap of policy
F: Follow-up 1-page
G: Subscribe to e-mail updates
H: Website location for more info
I: Co-branding with local agent
J: Provide call-in number
K: Highlight salient information
L: Bold top two to three information points
M: Box summarizing action steps
N: FAQs
O:
P:
Q: Kept open for focus group ideas
R:
S:

Design and Results:

Design: multifactorial 20-runs

Row	A	B	C	D	E	F	G	H	I	J	K	L	M	N	O	P	Q	R	S
1	+	+	−	−	+	+	+	+	−	+	−	+	−	−	−	−	+	+	−
2	+	−	−	+	+	+	+	−	+	−	+	−	−	−	−	+	+	−	+
3	−	−	+	+	+	+	−	+	−	+	−	−	−	−	+	+	−	+	+
4	−	+	+	+	+	−	+	−	+	−	−	−	−	+	+	−	+	+	−
5	+	+	+	+	−	+	−	+	−	−	−	−	+	+	−	+	+	−	−
6	+	+	+	−	+	−	+	−	−	−	−	+	+	−	+	+	−	−	+
7	+	+	−	+	−	+	−	−	−	−	+	+	−	+	+	−	−	+	+
8	+	−	+	−	+	−	−	−	−	+	+	−	+	+	−	−	+	+	+
9	−	+	−	+	−	−	−	−	+	+	−	+	+	−	−	+	+	+	+
10	+	−	+	−	−	−	−	+	+	−	+	+	−	−	+	+	+	+	−
11	−	+	−	−	−	−	+	+	−	+	+	−	−	+	+	+	+	−	+
12	+	−	−	−	−	+	+	−	+	+	−	−	+	+	+	+	−	+	−
13	−	−	−	−	+	+	−	+	+	−	−	+	+	+	+	−	+	−	+
14	−	−	−	+	+	−	+	+	−	−	+	+	+	+	−	+	−	+	−
15	−	−	+	+	−	+	+	−	−	+	+	+	+	−	+	−	+	−	−
16	−	+	+	−	+	+	−	−	+	+	+	+	−	+	−	+	−	−	−
17	+	+	−	+	+	−	−	+	+	+	+	−	+	−	+	−	−	−	−
18	+	−	+	+	−	−	+	+	+	+	−	+	−	+	−	−	−	−	+
19	−	+	+	−	−	+	+	+	+	−	+	−	+	−	−	−	−	+	+
20	−	−	−	−	−	−	−	−	−	−	−	−	−	−	−	−	−	−	−

Narrative:

This form of study uses the standard marketing approach of conjoint analysis, where design options (for a piece of hardware or service) are presented to potential consumers for subjective ranking and therefore feedback for redesign.

By dropping the approach into an orthogonal design, more information is revealed and organized with more structure. Specifically in this case up to 19 variants can be assessed rather than the "best" rough design or a few of them.

The survey instrument was designed as a phone call to current consumers to measure intent to retain or not (matching to real data over 90 days). The more important data was designed as actual retention two to three months later.

Data are due by June 2014 in this study currently underway.

In addition to the survey of hundreds of customers, a sample of 25 customers was invited to a focus group style session. All 20 packages were ranked (1-20) by each customer independently, as well as each of their cut points below which they not retain. This gave a separate analysis and a more objective statement as to which interventions caused retention, without ever asking.

7. Plant Productivity

Objective: increase production

Measurements: % of capacity

Test units: 4 production plants

Measurement error:

The only uncertainty was the maximum capacity, never met (with always some downtime).

Homogeneity:

The differences among plants are randomized out.

[individuals charts]   Month

Interventions:

A: Weekly all plant call on productivity tips
B: Preventive maintenance frequency higher
C: Noticeboard on productivity lessons learned
D: Q.C. production training
E: Cross shift overlap
F: Standardization across shift meetings
G: Hourly scoreboard
H: Feedback on defectives: daily data to
   production
I: Strict adherence to production schedule
J: Combine customer runs to reduce change
   over/set-up time
K: Short runs every other day only

Design and Results:

Design: multifactorial 12-run with foldover

Row	Plant	A	B	C	D	E	F	G	H	I	J	K	% of capacity
1	4	-	-	+	-	-	-	+	+	+	+	+	76
2	2	-	+	-	-	-	+	+	+	+	+	-	90
3	1	+	-	-	-	+	+	+	-	+	+	+	84
4	3	-	-	+	+	+	+	+	+	-	+	-	88
5	2	-	-	+	+	+	+	-	+	-	-	-	89
6	1	-	+	+	+	-	-	-	+	-	-	-	90
7	3	+	+	+	+	-	+	+	-	+	+	+	70
8	3	+	+	+	+	+	-	-	+	+	+	+	89
9	2	+	-	-	-	-	-	+	+	+	+	+	97
10	1	-	+	+	+	+	-	-	-	-	+	-	99
11	1	+	+	+	+	+	+	+	+	+	+	+	91
12	3	+	+	-	+	+	+	-	+	+	+	+	99
13	4	+	+	+	+	+	-	+	-	+	+	+	86
14	2	+	-	+	-	-	+	-	+	-	+	-	90
15	4	-	+	+	-	-	-	+	-	+	+	+	85
16	1	+	+	-	+	+	+	+	-	+	+	+	98
17	3	+	-	-	-	-	-	-	+	+	+	+	95
18	4	+	-	+	-	+	+	+	+	+	+	+	80
19	1	-	+	-	+	+	+	+	+	-	-	-	97
20	4	+	-	+	+	-	-	-	+	+	+	-	96
21	2	-	+	-	+	+	+	+	-	+	-	-	89
22	2	-	-	+	-	+	+	+	-	-	-	-	70
23	3	+	+	+	-	-	-	+	-	-	-	+	97
24	4	-	-	-	-	-	-	-	-	-	-	-	61
Effect		-0.7	5.7	0.7	1.7	2.3	6.0	0.8	-4.0	8.5	9.0	6.5	

Implementation: B, F, I, J, K

Narrative:

The 4th plant as a recent acquisition lagged the other 3. The 3 plants are numbered in the same order as illustrated in Figure 3.7. The following fractions will be found exceeding their previous average (by comparing to Figure 3.7, reproduced in this case's opening page).

Plant 1: 4/6
Plant 2: 5/6
Plant 3: 5/6

The test unit was a plant for a month, using the 4 plants. The gains during the test were seen from its start. Interactions IK and JK were -5.2 and -5.3 respectively. Though not quite significant, this is common, representing a diminishing return in two helpful interventions. Being smaller than the main intervention findings, these do not trump.

8. "Feet on the Street" Business to Business Sales.

Objective: Increase sales for reps. visiting new and current active business clients

Measurement: Sales points

Test units: "Feet on the street" reps

Measurement error:

The measurement was artificial sales points designed to incent a balance in priorities among products. In some artificial measurements of this type, conflict occurs when the numbers are not trusted. No one was talking about monthly commissions so the error check was skipped.

Homogeneity:

Prior year's sales was correlated so change in sales was used as the statistical design measurement.
Differences among reps were clear by eye so a simple homogeneity chart wasn't used. Non-homogeneity was later checked with a normal probability plot, looking for a split straight line.

Interventions:

A: Product specialist available to join meeting by phone
B: Larger new client emphasis
C: Soft skills seminar
D: Tag team weekly
E: Product seminar
F: Discount conditional on purchase patterns
G: Software enhancement
H: Contest
I: Daily vs. weekly goals
J: Leave behind material
K: Scripts
L: Enhanced demonstration
M: Send technical materials before meeting
N: Follow-up survey
O: Weekly stand-up meeting by phone on tips
P: Inside sales absorb "after meeting work"

Design and Results:
Design: Multifactorial 20-runs with foldover

Row	A	B	C	D	E	F	G	H	I	J	K	L	M	N	O	P	Q	R	S	change
1																				47.3
2																				37.4
3																				-14.2
4																				3.6
5																				29.7
6																				56.0
7																				63.6
8																				11.8
9																				12.6
10																				23.7
11																				-5.3
12																				12.6
13																				3.9
14																				6.4
15																				-14.8
16																				20.3
17																				49.6
18																				12.6
19																				0.8
20																				-16.5
21																				-17.8
22																				6.7
23																				49.3
24																				12.6
25																				-18.2
26																				-31.5
27																				3.5
28																				5.7
29																				34.7
30																				-4.5
31																				12.6
32																				18.4
33																				24.5
34																				26.1
35																				37.4
36																				13.0
37																				-9.4
38																				23.9
39																				17.4
40																				47.3
Effect	32.3	21.7	-2.0	-2.5	9.8	0.8	0.6	0.6	-0.5	-8.0	4.0	7.8	1.9	-0.8	2.1	2.8	1.2	-2.2	-1.2	

Designs and Results (*cont.*)

Normal Plot of Intervention Effects.

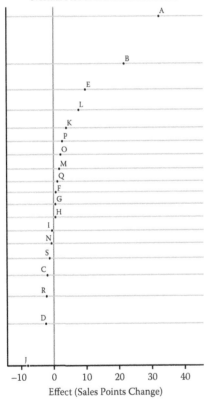

Effect (Sales Points Change)

The normal plot, since without its central, straight portion split, finds no aberrant results, indicating homogeneity pre and during testing.

2×2 on BL:

	+	4.7	32.7	Quadrant averages
L				
	−	3.3	18.6	

− B +

Cross diagonals:
BL= (32.7+3.3) / 2 - (4.7+18.6) / 2= 6.35

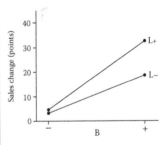

Implementation: A,B, E, L, not J

Prediction:

	Test average:	14.8
helps: A/2		16.1
B/2		10.8
E/2		4.9
L/2		3.9
B+L+/2		3.2
hurts: -J/2		4.0 +
		57.7 Sales change

Narrative:

Early decision centered on whether the "chemistry" and intuition in effective sales meetings would follow science and be statistically measurable. Inspection of the data and comparison to any of the supposedly more "tangible" cases will find no differences in data structure, using any statistical diagnostic or eyeball check.

The variation also reduces at the test and implementation.

[The individuals chart was prepared retrospectively.]

7

Simultaneous Design

... [a] wider inductive basis for our conclusions, without in any degree impairing their precision.

Sir Ronald Fisher*

7.1 SOLVING COMPLEX PROBLEMS SIMPLY

The quote means statistical designs give findings that are far more likely to be repeatable and therefore to implement close to the test's prediction. This advantage comes from testing in the real world rather than attempting to hold other conditions constant artificially. The "without impairing precision" means that each intervention is measured with the same statistical precision as if only a single intervention were tested. If it's accepted that testing more interventions brings advantage, it would follow that running more than one statistical design simultaneously would bring further advantage.

Every statistical design requires adaptation to fit the problem at hand. Little of this adaptation requires any new theoretical work. Simultaneous design solves a class of problem that arises often, using no new theory. Retailing provides a good example of the problem class. Consumers browse and buy products from more than one sales channel: stores, online, and outlets, perhaps also with catalog sales. Thus there are three to four sales channels. The problem is to decide which marketing, advertising, merchandising, sales, and delivery tactics will be optimal. It's also important

---

* Fisher, R.A. (1935, 1971, 2003). *The Design of Experiments*. (Reprinted, 2003.) Oxford, UK: Oxford University Press, p. 102.

not to cannibalize one channel for another. In retailing, solutions to this problem are known as cross-channel (or multichannel) optimization.

Where different call centers serve the same population of customers for different purposes, such as sales, service, and billing questions, the situation is similar. In the care management case, more complex variants of the same problem occur, when in addition to telephonic care by nurses, there are also pharmacists or house calls (by nurses or physicians), with the overall need to improve health, as before. That too is a similar problem. In telecommunications, different types of technicians (usually two or three) set up and later maintain the hardware and software for sophisticated entertainment/communications/Internet services, giving another example of the same problem class.

In education, more than one means is employed to educate students, such as classes, individual sessions, and through the use of educational products or self-study/homework approaches. The measurement aimed at is usually a grade at semester's end or longer term, or a pass/fail for a given examination or qualification.

In all of these examples, more than one system touches the output to be measured and improved.

7.2 SIMULTANEOUS DESIGN IDEA

When first faced with the multichannel problem, the immediate reaction, if using statistical design, is to run one on each system separately, in sequence. The business difficulty with this is that it takes too long and also gives no insight into how single-channel tactics might interact with tactics in another channel(s). Simultaneous design runs all channels at the same time with no technical disadvantages.

Because variation tends to reduce when running a statistical design, especially those involving people, a better time to run the other designs is when the first is running and all of the vice versas. This has a feel similar to when planting a sapling tree, where a single rope gives some protection against a gust of wind or a small animal knocking the tree over, but three ropes give considerably more protection. Each rope gives the others tension in a rigid triangular system. In statistical design, each tends to give the others reduced variation and everything in the analyses can only perform better. There is no intent to extend the sapling tree analogy to three designs (like ropes) as being optimal, but sets of three have been found common.

The obvious objection is that interactions across the designs will contaminate, leaving a mess that's impossible to untangle. At first glance this would certainly appear to be a problem. As the cases show, in fact it's not a problem and the illusion of one turns out to hide a welcome advantage. There are, however, some simple tricks to setting up simultaneous designs. These tricks would likely be missed if just copying how the designs look, without knowing what's inside them.

The next case, chosen because it makes simultaneous design appear quite obvious, also outlines the method. It is a case in which the tricks that make it work are almost intuitive and not much noticed. After that, the essential technical rules are provided. Then, a more complex real family of cases is set up, following exactly the same constructs as the first, but with the underlying constructs that make it work largely obscured from sight. Finally case fragments are given that suggest the flexibility of the method. The chapter concludes with a practical procedure that will apply to any problem of this type.

7.3 SCIENCE EDUCATION CASE

To increase academic performance in grade school science classes, the following interventions were assembled, some of which apply to classes, and others to individual students. The test units are therefore both classes and students. The two test unit types, rather than one, reveal the need for simultaneous design.

Early discussions covered the ethics of the test. The tendency in people tests is for increased performance during their course, therefore, teachers and administrators were supportive. Because they tried different tactics constantly, the experimental mode was always in play, as different teachers tried making change by varying the standard teaching model in tactical ways. The test would just put structure to it, and the facility to find out what worked. The final decision was to test only interventions incremental to the existing teaching model.

A later discussion raised the issue of whether the planned test was playing roulette with students and whether only volunteers should be invited to take part in the test. This would have biased the findings and revealed little about nonvolunteers. Instead, the decision was to allow the student interventions to be voluntary, while providing encouragement to adopt. Some of the class interventions followed the same approach, using intent to treat as always. The voluntarism will of course bias the test, as intent to treat always does, closer to the real world. It was also pointed out that inasmuch as a couple of interventions typically hurt, the test would stop any accidental roulette that had been occurring to date.

Overall, the statistical design was regarded as similar to standard practice, which left considerable flexibility for individual teachers, but structured it orthogonally.

7.4 DISCOVERY

Science teachers and staff met to discuss interventions and once more to prepare standards. The interventions are listed in Table 7.1. The list is split into class and student level interventions. Classes were divided into four levels (from basic to increasingly advanced), suited to different skill levels. These were built into the statistical design to find out what worked for all levels but without their differences contaminating the results.

TABLE 7.1

Science Education Interventions

Student Interventions:
A: Student leads off nonclass meetings
B: Early feedback
C: Mentor program
D: "Science in the news" homework
E: Essay questions as well as calculations
F: Online resource stressed
G: Choose personal project for completion by semester end, with guidance
H: Log study hours

TABLE 7.1 (continued)

Science Education Interventions

Class Interventions:

a	b	Level
–	–	1
+	–	2
–	+	3
+	+	4

c: Talk by industrial expert on uses of science
d: Prereading
e: Left blank for ab interaction (ab = –e, or e in foldover)
f: Student-friendly "math for science" sessions
g: More problems/calculation homework weekly
h: Scheduled question sessions
i: Enhanced awards program
j: Team building event
k: DVD monthly
l: Worked problems in each class
m: Extra workshops (optional)
n: Pop quizzes
o: Shorter presentation/more discussion

7.5 BASELINE DATA

Independent researchers had already calculated measurement precision using an internal validity check within the exam questions. This essentially measured each student twice and compared, then aggregated the whole for a final score, on a percentage scale. The data given in the research report were then combined with science grades for the 565 students planned for the statistical designs. This found the measurement error at 7.8% of variation in student scores.

Figure 7.1 shows the homogeneity check split into the four different class levels. The previous semester student science scores are plotted in random order. The chart excludes 23 students with low scores who were removed from the analysis of the study. Four slight exceptions were favored over unbalancing the design.

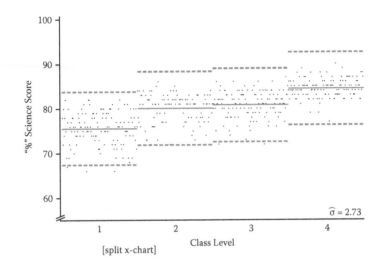

FIGURE 7.1
Homogeneity of science exam scores prestudy.

7.6 SIMULTANEOUS STATISTICAL DESIGNS FOR SCIENCE CLASSES

The two designs, for 512 students, are shown in Figure 7.2a and b. The classes were randomized to the class design and the students within each class were randomized to the student design. Thirty extra students were

Row	A	B	C	D	E	F	G	H
1	−	−	−	−	−	−	−	−
2	+	−	−	−	+	−	+	+
3	−	+	−	−	+	+	+	−
4	+	+	−	−	−	+	−	+
5	−	−	+	−	+	+	−	+
6	+	−	+	−	−	+	+	−
7	−	+	+	−	−	−	+	+
8	+	+	+	−	+	−	−	−
9	−	−	−	+	−	+	+	+
10	+	−	−	+	+	+	−	−
11	−	+	−	+	+	−	−	+
12	+	+	−	+	−	−	+	−
13	−	−	+	+	+	−	+	−
14	+	−	+	+	−	−	−	+
15	−	+	+	+	−	+	−	−
16	+	+	+	+	+	+	+	+

Student Design

(a)

FIGURE 7.2
(a) Student design and (b) class design. *(continued)*

Row	a	b	c	d	e	f	g	h	i	j	k	l	m	n	o	
1	+	-	-	-	+	-	-	+	+	-	+	-	+	+	+	
2	+	+	-	-	-	+	-	-	+	+	-	+	-	+	+	
3	+	+	+	-	-	-	+	-	-	+	+	-	+	-	+	
4	+	+	+	+	-	-	-	+	-	-	+	+	-	+	-	
5	-	+	+	+	+	-	-	-	+	-	-	+	+	-	+	
6	+	-	+	+	+	+	-	-	-	+	-	-	+	+	-	
7	-	+	-	+	+	+	+	-	-	-	+	-	-	+	+	
8	+	-	+	-	+	+	+	+	-	-	-	+	-	-	+	
9	+	+	-	+	-	+	+	+	+	-	-	-	+	-	-	
10	-	+	+	-	+	-	+	+	+	+	-	-	-	+	-	
11	-	-	+	+	-	+	-	+	+	+	+	-	-	-	+	
12	+	-	-	+	+	-	+	-	+	+	+	+	-	-	-	
13	-	+	-	-	+	+	-	+	-	+	+	+	+	-	-	
14	-	-	+	-	-	+	+	-	+	-	+	+	+	+	-	
15	-	-	-	+	-	-	+	+	-	+	-	+	+	+	+	
16	-	-	-	-	-	-	-	-	-	-	-	-	-	-	-	
17	-	+	+	+	-	+	+	-	-	+	-	+	-	-	-	
18	-	-	+	+	+	-	+	+	-	-	+	-	+	-	-	
19	-	-	-	+	+	+	-	+	+	-	-	+	-	+	-	
20	-	-	-	-	+	+	+	-	+	+	-	-	+	-	+	
21	+	-	-	-	-	+	+	+	-	+	+	-	-	+	-	
22	-	+	-	-	-	-	+	+	+	-	+	+	-	-	+	
23	+	-	+	-	-	-	-	+	+	+	-	+	+	-	-	
24	-	+	-	+	-	-	-	-	+	+	+	-	+	+	-	
25	-	-	+	-	+	-	-	-	-	+	+	+	-	+	+	
26	+	-	-	+	-	+	-	-	-	-	+	+	+	-	+	
27	+	+	-	-	+	-	+	-	-	-	-	+	+	+	-	
28	-	+	+	-	-	+	-	+	-	-	-	-	+	+	+	
29	+	-	+	+	-	-	+	-	+	-	-	-	-	+	+	
30	+	+	-	+	+	-	-	+	-	+	-	-	-	-	+	
31	+	+	+	-	+	+	-	-	+	-	+	-	-	-	-	
32	+	+	+	+	+	+	+	+	+	+	+	+	+	+	+	Class Design

(b)

FIGURE 7.2 (continued)
(a) Student design and (b) class design.

included, as partial replicates, also allocated randomly. Of course, students cannot be randomly allocated to classes, already being constrained by class level. The scheduling procedure used a method that had evolved over the years, taking into account several criteria. At any rate, as with the care management case, there is no need to randomize students to classes but it's often asked about or suggested.

Technically, the two designs are a fractional factorial for the student design (of the same type as used in the retail case) and a multifactorial 16 row with foldover, making 32 rows. The foldover is clear by looking at rows 16 and 32, which are reversed, as are all other row pairs reading up (e.g., rows 31 and 15, 30, 14, etc.).

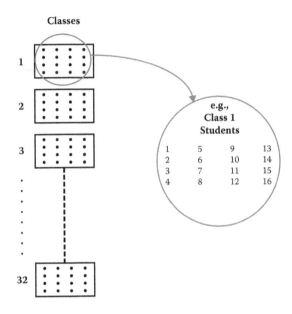

FIGURE 7.3
Classes contain student designs.

The important construct that is self-evident in this particular case is that each row of the class design contains a student design (Figure 7.3). This feature is found in all simultaneous designs, although usually less obviously to the eye.

The pair of designs is further clarified by the final data shown in Figure 7.4. These data are the pre scores and the change during the semester the study was run. (Thus the study score is revealed by addition.) Reading down the page gives the 32-row class design, reading across gives the 16-row student design in columns. These were the data used in the homogeneity check. The row and column averages are now ready for simple analysis.

Change in score is used throughout (i.e., test minus pre). Starting with the student design, Figure 7.5 shows the analysis, including aliased pair interactions. The data are the column averages in Figure 7.4, with the class intervention effects averaged out: essentially cancelled, neutralized, or "gone" for now.

No intervention was therefore significant in the dry run. In the test, clearly G and one aliased interaction are significant, undiluted by the dry run.

Figure 7.6 shows the class design analysis, this time with student intervention effects eliminated. Significance calculations place these results in clearer context.

CALCULATION (METHODS 2 AND 3)

With 512 students needed for the study, this left 30 partial replicates and therefore a useful direct measure of student-to-student variation in regard to change in score. After randomizing for the designs, this gave an average range (for the 30 student pairs) of 2.17 percentage points. (Strictly speaking, these are not percentages but rather scores on a scale of 0 to 100; traditionally they were thought of as percentages.) This then gave:

$$\hat{\sigma} = 2.17/1.128 = 1.92$$

Therefore the noise is $\pm t\hat{\sigma}\sqrt{(4/N)}$, where t has 30 degrees of freedom (from the 30 partial replicates): $30 \times (2 - 1) = 30$. N is the 512 students. Noise $= 2.042 \times 1.92 \times \sqrt{(4/512)} = 0.347$.

The dry run analysis is also shown, but this has a different noise level, being prior to the statistical design.

$\hat{\sigma}_{pre} = 2.73$ (Figure 7.1 or source data) therefore the noise for the dry run uses the same formula but with degrees of freedom coming from all the study students where $t = 1.96$. Therefore noise $= 1.96 \times 2.73 \times \sqrt{(4/512)} = 0.473$.

The different class levels are significantly different in the dry run, as expected, but serve only to remove those differences from other intervention effects. Figure 7.7 shows the likely pair interactions for both designs under effects heredity.

There are also 120 pair interactions across the designs: Aa, Ab, Ac, and so on, all the way to Ha, Hb, ..., Ho. The significant ones are shown in Figure 7.8 and plotted in Figure 7.9. The data rearranged in Figure 7.8 are easy but tedious to assemble from the earlier source data (Figure 7.4).

	Pre																Change																Average	
Row	1	2	3	4	5	6	7	8	9	10	11	12	13	14	15	16	1	2	3	4	5	6	7	8	9	10	11	12	13	14	15	16	Pre	Change
1	77	83	84	78	81	74	80	81	83	80	85	74	81	86	80	83	6	4	-1	1	-2	4	4	1	2	0	-1	-4	2	-1	3	0	80.63	1.31
2	84	79	87	87	84	83	83	85	84	85	83	88	84	86	82	82	2	1	-1	2	1	3	3	1	0	-1	1	-1	2	1	3	3	84.25	1.19
3	82	83	88	87	85	82	83	85	85	88	83	83	86	85	85	84	1	5	1	0	0	3	0	-1	3	-1	3	2	2	0	2	2	84.63	1.25
4	86	83	85	84	85	86	82	84	79	85	82	83	85	83	83	84	0	3	1	-1	-1	0	1	2	3	1	0	3	0	-1	3	0	83.69	0.81
5	78	79	87	75	80	85	80	83	81	76	80	80	80	82	82	84	3	3	-2	-1		-6	1	2	2	1	0	3	0	2	-1	1	80.69	0.81
6	83	76	76	80	81	73	80	83	80	80	79	80	79	72	73	73	2	3	1	1	4	1	1	2	4	2	2	1	2	6	-1	1	78.25	1.44
7	75	84	79	82	82	82	82	83	83	81	83	77	83	76	82	81	3	0	4	3	0	1	3	1	4	2	2	3	4	0	0	4	80.69	1.88
8	82	77	75	85	81	83	77	79	81	75	82	82	81	81	72	76	3	-2	5	4	2	0	4	1	0	2	2	0	0	0	4	-2	79.35	1.63
9	86	87	84	86	83	84	83	85	83	87	82	77	84	83	82	85	0	1	5	4	0	6	1	1	0	1	4	1	0	2	0	0	83.75	1.06
10	83	84	84	82	78	75	78	81	79	76	76	80	81	83	80	83	1	4	-2	2	0	-1	-4	2	4	-2	4	1	4	-1	1	3	80.56	1.81
11	77	73	75	75	77	78	73	81	75	74	73	74	76	74	76	79	1	6	5	2	0	3	3	-1	2	1	3	4	0	3	2	4	75.44	0.88
12	77	76	84	78	79	82	76	84	77	82	79	81	80	81	81	80	4	3	5	0	2	4	3	-1	0	2	3	0	1	0	-1	4	79.81	1.94
13	76	79	80	77	84	78	74	72	82	83	73	83	79	75	82	83	4	2	2	5	1	4	1	-1	2	1	2	4	1	2	5	4	80.00	2.25
14	75	76	66	79	71	75	80	72	66	75	75	79	79	75	74	74	-2	0	0	5	1	4	1	-1	3	4	5	4	0	2	5	-2	74.44	1.56
15	72	74	76	79	72	78	80	72	70	78	75	77	74	80	71	76	0	4	4	3	8	3	2	0	3	3	-1	1	2	2	2	2	75.25	2.50
16	74	80	69	75	76	77	76	78	79	75	75	69	71	72	71	71	2	2	2	7	-2	4	-1	-3	5	0	-1	7	3	-1	5	2	74.25	2.00
17	82	83	80	86	84	84	81	78	79	80	77	78	76	81	77	72	2	4	2	2	-2	4	4	0	1	2	-7	0	3	3	-1	3	79.81	2.19
18	76	80	69	81	81	78	78	76	69	77	77	78	74	71	76	75	4	3	4	0	2	3	0	0	1	2	-1	0	6	3	0	6	75.56	1.44
19	73	78	81	80	72	76	80	71	79	79	82	76	80	77	82	72	4	3	0	-2	3	2	2	0	3	1	-1	6	4	5	0	6	77.38	2.00
20	72	76	80	68	80	81	82	73	69	76	79	78	75	80	70	78	-2	3	-3	0	1	5	-2	1	2	3	-3	0	5	2	1	1	76.06	0.63
21	85	81	77	78	85	81	82	80	83	81	83	76	84	81	81	81	2	3	-1	1	0	1	-1	3	1	2	1	1	0	-1	1	-4	80.50	0.63
22	81	84	82	80	84	82	80	80	80	81	78	77	80	86	84	79	2	-3	3	-3	2	0	3	-2	-2	0	2	1	0	-1	-2	3	82.13	0.13
23	82	84	77	80	80	81	83	82	77	83	83	82	84	81	78	84	2	2	3	1	1	0	0	3	2	0	2	0	1	2	2	1	80.81	1.06
24	77	79	81	78	80	84	81	81	79	82	78	81	84	84	84	81	6	9	4	-2	0	0	5	0	3	-1	0	-1	-2	0	-4	0	81.31	0.69
25	79	78	77	81	75	77	69	75	76	81	83	74	76	82	67	80	3	-1	2	-2	-1	3	2	2	2	1	-1	3	-1	0	5	-2	76.50	1.00
26	82	80	82	82	83	85	79	81	83	81	84	81	76	81	82	78	1	1	0	4	0	-1	2	1	0	0	3	-1	4	0	0	2	81.25	0.44
27	87	87	87	85	87	85	85	83	85	85	86	84	84	85	84	85	-1	3	4	0	-1	-1	2	2	1	1	-2	0	0	1	0	3	85.25	0.44
28	82	84	81	85	82	78	78	78	80	82	77	83	84	82	80	82	1	2	0	0	-2	3	3	1	1	2	-4	2	0	7	2	1	81.19	1.56
29	80	80	80	82	73	81	84	82	78	79	90	79	76	83	83	80	1	2	1	2	8	2	0	2	2	2	0	2	2	0	2	2	80.00	1.81
30	82	88	80	83	84	82	84	82	83	82	82	86	84	85	87	86	4	1	2	6	-2	3	2	4	3	0	-2	-6	4	-9	1	-1	84.69	-0.19
31	81	84	84	81	85	89	86	82	85	82	85	85	85	86	84	87	-2	0	2	1	1	-9	1	-1	0	-2	-8	-1	2	-9	3	-9	84.44	0.31
32	89	84	82	85	83	85	84	82	85	88	82	84	85	86	87	83	-2	1	1	2	4	2	2	-1	0	-2	2	2	2	0	-9	-9	84.88	0.25
Average	79.91	80.94	80.19	80.69	80.22	80.75	80.34	80.44	79.28	80.56	80.47	79.59	80.56	80.41	79.44	79.94	1.63	1.97	1.34	1.34	1.19	1.44	1.09	0.78	1.59	1.03	0.28	0.84	1.78	0.56	1.09	1.69	79.91	1.69

Class Row / *Student Row*

FIGURE 7.4

Test data for the two simultaneous designs.

Row	A	B	C	D	E	F	G	H	FH/DG/CE/AB	FG/DH/BE/AC	EF/CH/BG/AD	GH/DF/BC/AE	BH/CG/DE/AF	EH/CF/BD/AG	EG/BF/CD/AH	pre	change
1	−	−	−	−	−	−	−	−	+	+	+	+	+	+	+	79.91	1.63
2	+	−	−	−	+	−	+	+	−	−	−	+	−	+	+	80.94	1.97
3	−	+	−	−	+	+	+	−	−	+	+	−	−	−	+	80.19	1.34
4	+	+	−	−	−	+	−	+	+	−	−	−	+	−	+	80.69	1.34
5	−	−	+	−	+	+	−	+	+	−	+	−	−	+	−	80.22	1.19
6	+	−	+	−	−	+	+	−	−	+	−	−	+	+	−	80.75	1.44
7	−	+	+	−	−	−	+	+	−	−	+	+	+	−	−	80.34	1.09
8	+	+	+	−	+	−	−	−	+	+	−	+	−	−	−	80.44	0.78
9	−	−	−	+	−	+	+	+	+	+	−	+	−	−	−	79.28	1.59
10	+	−	−	+	+	+	−	−	−	−	+	+	+	−	−	80.56	1.03
11	−	+	−	+	+	−	−	+	−	+	−	−	+	+	−	80.47	0.28
12	+	+	−	+	−	−	+	−	+	−	+	−	−	+	−	79.59	0.84
13	−	−	+	+	+	−	+	−	+	−	−	−	+	−	+	80.56	1.78
14	+	−	+	+	−	−	−	+	−	+	+	−	−	−	+	80.41	0.56
15	−	+	+	+	−	+	−	−	−	−	−	+	−	+	+	79.44	1.09
16	+	+	+	+	+	+	+	+	+	+	+	+	+	+	+	79.94	1.69
Change:																	
Av+	1.207	1.059	1.203	1.109	1.258	1.340	1.469	1.215	1.355	1.164	1.172	1.359	1.285	1.266	1.426		
Av−	1.250	1.398	1.254	1.348	1.199	1.117	0.988	1.242	1.102	1.293	1.285	1.098	1.172	1.191	1.031		
Effect	−0.043	−0.340	−0.051	−0.238	0.059	0.223	0.480	−0.027	0.254	−0.129	−0.113	0.262	0.113	0.074	0.395		
Noise	0.347																
Dry Run:																	
Av+	80.41	80.14	80.26	80.03	80.40	80.13	80.20	80.29	80.08	80.17	80.14	80.10	80.40	80.16	80.26		
Av−	80.05	80.33	80.20	80.43	80.10	80.33	80.27	80.18	80.39	80.29	80.32	80.40	80.06	80.31	80.21		
Effect	0.36	−0.19	0.06	−0.40	0.36	−0.20	−0.07	0.11	−0.31	−0.12	−0.18	−0.25	0.34	−0.15	0.05		
Noise	0.473																

FIGURE 7.5
Student design results and analysis. The design was generated using E = ABC, F = BCD, G = ABD, H = ACD.

FIGURE 7.6
Class design results and analysis.

Row data (response columns):

Row	Pre	Change
1	80.63	1.31
2	84.25	1.19
3	84.63	1.25
4	83.69	0.81
5	80.69	1.44
6	78.25	1.44
7	80.69	1.88
8	79.38	1.63
9	83.75	1.06
10	80.56	1.81
11	75.44	0.88
12	79.81	1.94
13	80.00	2.25
14	74.44	1.56
15	75.25	2.50
16	74.25	2.00
17	79.81	2.19
18	75.56	1.44
19	77.38	2.00
20	76.06	0.63
21	80.50	0.63
22	82.13	0.13
23	80.81	1.06
24	81.31	0.69
25	76.50	1.00
26	81.25	1.06
27	85.25	0.44
28	81.19	1.56
29	80.00	1.81
30	84.69	-0.19
31	84.44	0.31
32	84.88	0.25

Change:

	a	b	c	d	e	f	g	h	i	j	k	l	m	n	o	Pre
Av+	1.00	1.03	1.24	1.29	1.18	1.28	1.32	1.20	1.09	1.22	1.09	1.30	1.21	1.30	1.11	80.23
Av-	1.46	1.43	1.22	1.17	1.27	1.18	1.14	1.26	1.37	1.24	1.37	1.16	1.25	1.15	1.35	
Effect	-0.46	-0.40	0.02	0.11	-0.09	0.11	0.18	-0.07	-0.28	-0.02	-0.29	0.14	-0.04	0.15	-0.25	1.23
Noise	0.347															

Dry Run:

	a	b	c	d	e	f	g	h	i	j	k	l	m	n	o
Av+	82.26	82.62	80.02	80.15	80.30	80.11	80.17	80.36	80.41	80.17	80.37	80.34	80.25	80.30	80.48
Av-	78.20	77.84	80.45	80.31	80.17	80.36	80.30	80.10	80.05	80.29	80.10	80.12	80.22	80.17	79.99
Effect	4.06	4.78	-0.43	-0.16	0.13	-0.25	-0.13	0.26	0.36	-0.12	0.27	0.22	0.03	0.13	0.49
Noise	0.473														

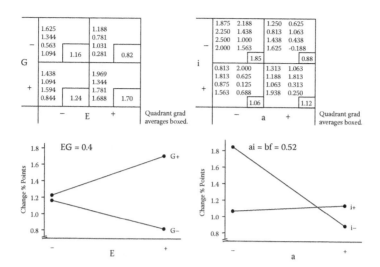

FIGURE 7.7

Change analysis: significant interactions.

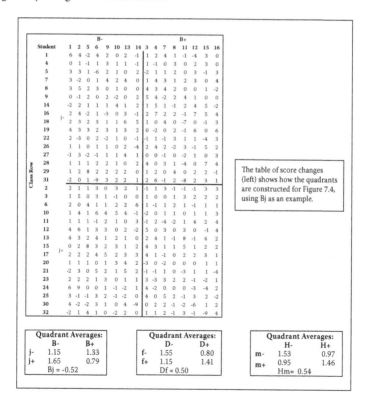

FIGURE 7.8

2×2 Projections of original data onto significant interactions across designs as 2×2 tables.

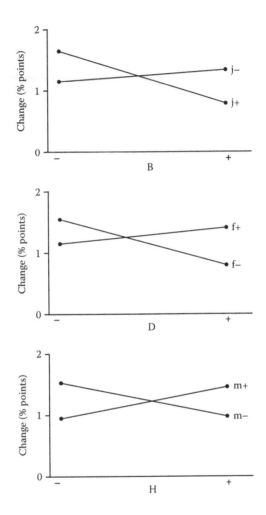

FIGURE 7.9
Pair interactions across student and class designs.

7.7 PAIR INTERACTIONS ACROSS DESIGNS
AND AN EASIER ANALYSIS

The 120 cross-design interactions, including the three plotted in Figure 7.9 are all unaliased. This is easy to see because all columns in each design are unique therefore so too will be any 2×2 across them. Therefore the interactions across designs that appear at first to be a problem in fact bring an advantage. They are all independent. That makes them more attractive than the interactions within each design, which are aliased with several others.

An easier way to analyze everything at once is to prepare both designs with all 142 pair interactions into a 512-row matrix and run a regression. All 23 interventions, the 15 aliased sets of pair interactions in the class design, the 7 aliased sets in the student design, and the 120 unaliased pair interactions across the two designs will be found mutually orthogonal.* Then the entire analysis of everything so far can be done in a single run. The dry run needs another.

7.8 FINDINGS

The results are self-explanatory, ending up rather simply. Choosing a personal project for completion by semester's end, with guidance from the science teacher (G), increased change in score by 0.48% points, enhanced almost as much by essay questions as well as calculations (E). ai and bf being aliased is troubling. They resolve shortly.

The cross-interactions indicate that the team building event (j) and no early feedback (B) help, and that the science in the news homework (D) and student-friendly math (f) are best dropped. This trumps the earlier result that (i) helped certain levels, which was aliased. Logging study hours (H) and extra workshop options (m) are best also dropped.

The findings make use of heredity of effects. The class-level interactions are of less interest than the main purpose: to split the class differences in order that other interventions would not be contaminated. This leads to an implementation plan of

E, G, j

---

* Statisticians will find the designs mutually orthogonal, by running a correlation matrix (in the interventions plus interactions with all $32 \times 16 = 512$ rows), whose non-diagonal entries are all found to be zero.

The predicted improvement, using the usual calculation becomes:

G	0.48
EG	0.40
Bj	0.52
Df	0.50
Hm	0.54
Total	2.44

The improvement is therefore roughly predicted at 2.44/2 = 1.22 points over the test average.

This sounds slight but translates to a noticeable increase in percentile against national performance. Those data were also analyzed with about a 5 to 1 leverage. It's hard to predict exactly but around 5 × 1.22 points increase in national percentiles would be in order. In the year following, there was an 8.4% increase against national percentiles.

7.9 RULES FOR SIMULTANEOUS DESIGNS

The simple rule is that each design be contained in the other(s), as was obvious in the science education case. The other rule that remains essential is that test units (here, both students and classes) be randomized to the test designs. There is no requirement that students be randomized to classes, nor is that possible.

7.10 GENERAL MULTICHANNEL OPTIMIZATION CASE

In the science education case, the student design was entirely contained within each class, with the 32 classes forming the class design. Often, each design being contained within each other's design is not as obvious. Take, for example, three consumer sales channels: original call center, transfer call center, and direct marketing packages sent to consumers' homes. Groups of potential customers, some of whom buy, form the test units as depicted in Figure 7.10. It is now clear that each design is contained within the other two but it's not known where a specific consumer will fall until it

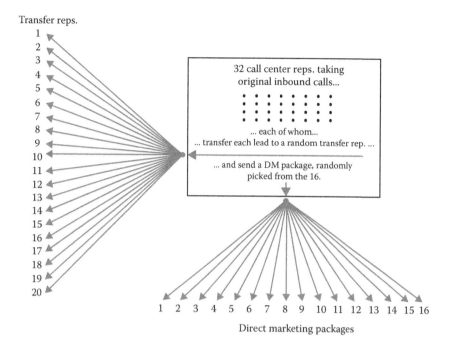

Transfer reps.

FIGURE 7.10
How the set of three simultaneous designs generally works.

happens. It turns out then that the structure is exactly the same as for the science education designs. The only differences are it's harder to see except with Figure 7.10 (which is usually sketched on a scrap of paper to make sure) and there are three simultaneous designs rather than two.

The way the thing works is that inbound calls from potential consumers are placed to the original call center. After purchasing (or not) the main product, the consumers become leads for the peripheral products. The two channels that influence peripheral purchases are transfer reps (in a different call center miles away from the original one) and peripheral product packages sent direct to the consumer. The original reps already have influenced future peripheral purchases in ways the simultaneous analysis will later reveal.

The rule to allow the three simultaneous designs to work is that each be fully contained within the other. It's obvious from Figure 7.10 this is so, because a line can be drawn from each original rep to each transfer rep and then on to each direct marketing package and all the vice versas. It takes a little while for the whole three-dimensional matrix to populate, needing at least $32 \times 20 \times 16 = 10,240$ transfer calls. This is not a high volume, especially in the month that tests of this type usually run.

The simple question to ask to ensure that simultaneous designs will provide a valid test is: "Could each prospect go to any original rep and then (if a buyer) to any transfer rep and receive any marketing package?" In general, the question to ask is: "Could any incoming subject or transaction follow a path through any of the test units for all simultaneous designs?" The reps and packages are randomized to each of their designs.

Notice that the fundamental unit of the set of designs is below the level of each test unit, at the individual level. It is this that makes the designs work. In cases where a test unit in one channel is dedicated to one or a few test units in another channel(s) then a single design is used, with the composite test unit as that "one to few." A simultaneous design would be ill-advised. Flows (or touches) from one design to the other(s) may be haphazard and needn't be random, as long as each design is randomized to its own test units.

7.11 SIMULTANEOUS DESIGN PROCEDURE

An easy way to prepare any statistical design is to research and list interventions first, then ask for each, "What is the test unit to control each intervention?" If the answer is the same for all interventions then it's a single design. If the answers differ then the list can be rearranged into sublists of interventions answered the same. Those sublists form the interventions for each of the set of simultaneous designs. Then the flow of each design to each other, by test units is checked. If the flow is restricted (e.g., one inside rep is dedicated to a few field people) then the 1: few becomes a composite test unit for a single design. If the flow is "all to all" for every design to every other one then the simultaneous designs work well. Figure 7.11 illustrates this for any pair of test units encountered (whether "flows" or "touches"), readily applied to more than two designs by checking every pair in this way.

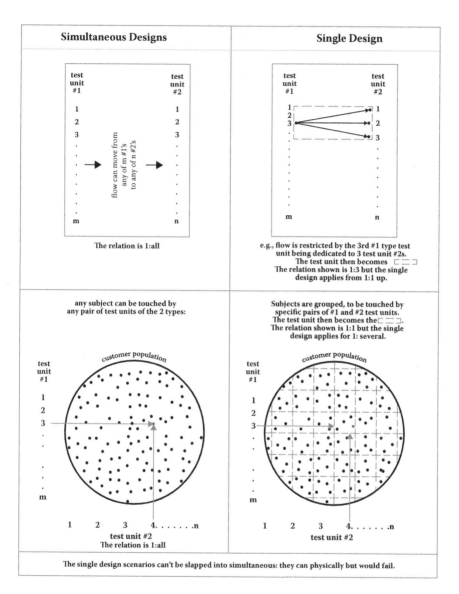

FIGURE 7.11

Simultaneous versus single designs.

8

Scientific Method, Randomization, and Improvement Strategies

I learned ... that innovation is a very difficult thing in the real world.

Richard Feynman[*]

8.1 SIMPLICITY OF THE SCIENTIFIC METHOD

Many nontechnical people have been found previously intimidated by science, mathematics, and statistical methods when starting work on statistical design and control to solve a business problem. It's been reassuring to point out that difficult problems may indeed require difficult mathematics, but most problems are easier. Statistical design and control are now placed in their scientific context, as a way to stimulate and accelerate scientific learning by everyone, for business competitiveness.

A toddler, given his first balloon, after awhile released it. Studying the balloon for a full two minutes as it became a speck in the sky, the toddler burst into tears. All future balloons were held onto firmly.

What happened there was the scientific method before speech was fully formed. It was innate. As the toddler had over the years learned a rule (also called a *model*) that objects fell to the ground when released, the initial <u>deduction</u> was that the balloon would too. When the balloon slowly became a speck in the sky, a crude measurement by eye was adequate and the toddler realized that the balloon would not be coming down soon in any playable form. The new data (assuming a randomly selected balloon,

---
[*] Feynman, R.P. (1985). *Surely You're Joking, Mr. Feynman!* New York: Norton.

with a bunch of those even better) led by <u>induction</u> to a new rule that balloons go up and had better be held onto in the future ...

... and learning occurred: "Some of What goes up must come down," became the new model.

Statistical design and control has been found to harness that innate way of learning, so that something can be innovated or improved.

The formal definitions of induction and deduction (the scientific method) are less direct than the toddler's view of things. *Induction* is the generalization of the specific and *deduction* draws the specific from the general, but it's not so clear what that means. The toddler translated and 30 years later recalled the event noting the exact reasoning summarized above.

Theory means a model, preferably mathematical, that stands up well to continuing data, subject to updates. The toddler's model was a simple rule. When the initial model "failed" with the balloon test, it wasn't proven wrong; it was just incomplete, therefore it was upgraded. The toddler's learning is identical to competitive advantage from innovation. Statistical design and control are founded firmly in theory, meaning they will work well in future application in the real industrial world.

8.2 SCIENTIFIC METHOD WITH STATISTICAL DESIGN AND CONTROL

The research, to list interventions (e.g., a special pharmacy review), sets up hypotheses (coming from observation of past experience or data). The randomized statistical design, for now assuming some interventions will work (deduction by each suggester), leads (with its analysis) to finding which ones do. By collective induction the nurses explained why, leading to a new implementation model.

Then the cycle continued under a new deduction that the handful of helpful interventions ought to help by 24%+, the initial implementation data at only about 12% reduction found a gap and more induction about a productivity clash followed; then the model was upgraded again by maintaining the handful of helpful interventions plus a change in productivity management. The final deduction that this would work was confirmed by data at 17%, still decreasing. Strictly speaking the case ought to have gone on to another cycle to squeeze out the remaining shortfall from the exact prediction. The reality though was that this one had succeeded and sustained whereas other priorities were now more pressing, comparatively. Also the prediction has uncertainty so it's really 24% give or take. Because the hospitalization rate was still steadily reducing it was better to let nature take its course for awhile longer. Trying to force a faster rate of reduction could have unsettled what was already working. With statistical design, 20+ interventions finds approximately the optimum among 2^{20+} = a million+ possibilities, therefore the rate of acceleration of learning (competitiveness) can be appreciated.

8.3 RANDOMIZATION DISTRIBUTION

In the care management case, the pretest admit rates were given in Figure 1.6, randomly ordered. In regard to any intervention, the randomization allocates 10 nurses, with the other 10 to its counterfactual. This was just one of many randomizations that could have occurred. In fact there are 184,756 ways to split the 20 nurses into 10 for the intervention versus 10 to its counterfactual.[*] Figure 8.1 shows 10,000 of these random

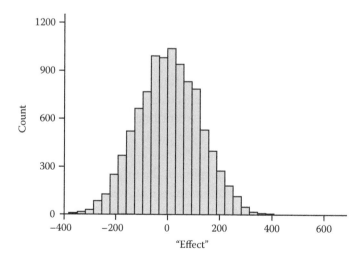

FIGURE 8.1
The randomization/distribution: potential pretest biases.

---

[*] $^{20}C_{10} = 20!/(10!10!) = 184{,}756$. The randomization distribution is known, perhaps ironically in view of its name, as distribution-free. It neither assumes nor requires normality of the original data.

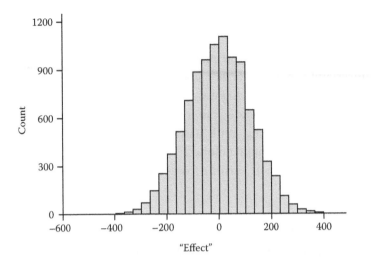

FIGURE 8.2
The randomization distribution: random splits on test data.

contrasts, still on the pretest data. These are possible "effects" of an intervention before it runs, in other words the possible pretest biases. Clearly they average about zero and the 5% of biases in the tails are significant. This gives a clearer view of why pretest bias is always inherent and tends to arise significantly in one intervention of every 20 tested over the long run on several statistical designs. This isn't a problem but underlines the need to adjust for pretest bias or at least take it into account using common sense.*

Figure 8.2 shows the test data in the same way. That is, it performs 10,000 random splits and calculates the "effect" of each. The 5% in the tails (2.5% either side) cuts off at about ±228. Accordingly, any interventions outside the cutoffs are significant. R, for example, is therefore significant with an effect of –261.4, "confirmed" by its dry run at –22.0.

Counting the 10,000 random splits in the original data summarized in Figure 8.2 found 100 of the random splits had "effects" more negative than –261.4. (As luck would have it, the first run of the random splits for publication produced this 100 as a nice round number!) So the actual effect at –261.4 is surpassed (more negatively) by 100 of the 10,000 random

---

* A cable technician hearing consternation that randomization didn't eliminate bias (in a statistical design to reduce install time and callbacks) asked the group: "Why, when ducks fly in a V, is one side of the V longer than the other?" No one knew so he added: "More ducks in it." His wit captured the phenomenon perfectly.

"effects," meaning 100/10,000 = 0.01 (akin to 1%). Because we're trying to reduce admit rates, the one-sided view is appropriate. (If we were trying to find anything that influenced admit rate, good or bad, then the two-sided calculation found also that 97 other random effects exceeded +261.4, adding to the one-sided 100: 197/10,000 = 0.0197. This clears the 0.05 significance hurdle.) Other methods of assessing significance will approximate this one, usually quite well.

No assumption of a bell-shaped curve (or any shaped curve) was used, only the raw data taken as they were found. This has introduced the *randomization distribution*. In the original development of statistical design theory [4], the randomization distribution was used to assess significance. What's important is that it does so directly with no mathematics or assumptions.

In many cases the randomization distribution will fall very close to a bell-shaped curve. Its pretest form reveals exactly the pretest biases that can be expected. Were it to be argued that randomization ought to eliminate significant bias, data would need to be produced that gave no tails when subjected to this procedure. That is impossible.

The randomization distribution for the retail sales data yielded a one-sided for the largest effect (A at 77.4) as 0.005. The traditional analysis again came close. This used the 16 averages for the store pairs, for 12,870 possible random splits.* All were run, with 65 exceeding 77.4. So 65/12780 = 0.005, again significant.

In practice the randomization distribution isn't always used, rather the standard methods and software. It mainly affords a clearer understanding of the randomization device, the pretest biases, and how orthogonal design uses randomization. It provides a simple, practical, direct assessment of significance where mathematical model assumptions are violated.

8.4 RANDOMIZATION DEVICE

Life does not come at us randomly. Other than in casinos, most have no experience of what the randomization device really accomplishes or how it works. Mathematicians often suspect that variables known to affect the

---

* The 32 original data points might be preferred but create $^{32}C_{16} = 601,080,390$ possible random splits and an impossible calculation volume; the simpler case does well enough.

statistical design's measurement could affect the test's findings. Or that some other influence could cause a particular result, other than the intervention. Cases have already shown that neither occurs. Randomization understandably may be hard to trust in general without extensive experience in running statistical designs or trying the exercises throughout this book.

The clinical care and retail cases both showed that all other variables in play (even with known strong effects on the outcome measurement following the objective) did not change the findings. This will always be the case, or defused by the dry run. Those cases also showed that the chance of something other than each intervention causing its result were very near zero: effectively not possible in practice provided the design is of a decent size.

It's also been shown how randomization makes otherwise unusable measurement systems usable (provided in statistical control) and is not upset by strong influences such as seasonality, step changes, and economic conditions, local or larger, always provided good statistical design is employed. The reasonable fear though is that in one case or another, randomization could fail and deliver a finding that was in fact caused by some other variable(s). Let's see from where that "could" might have come.

In studying mathematics, it might be asked which of the following "could" be random sequences:

(a) 7 3 10 4 2 5 8 9 1 6
(b) 1 2 3 4 5 6 7 8 9 10
(c) 3 8 4 2 10 7 6 9 5 1

and you're supposed to answer that they all "could."

Notice, however, that the second sequence is obviously unlikely in a real randomization (as are all three). It has a 1/10! chance of being drawn, which is once every 3,628,800 times you randomize, in the long run. Therefore it will happen 0.0000276% of the time. The other two sequences are of course equally unlikely but randomizations with strange patterns such as precisely 1–10 that "could" thwart a statistical design are not expected. (If a chemical process were prone to drifting over time or a person were more fatigued at the end of a study than its start then the "random" sequence 1–10 would be troublesome.) As unlikely is a sequence that coincided precisely (or even roughly) with an intervention in any setting, not just one that changed with time.

To drive home this missed point, mathematicians are sometimes given a story in which lots of monkeys typing at random would eventually type the complete works of Shakespeare. The idea is supposedly that the Shakespearean opus is as likely as any other random sequence. Although true, a couple of problems arise. First, mathematicians are comfortable with infinite series and other things that use infinity such as (in the monkey story) infinite time. Most industrial problems have to be solved much faster. Second, the story uses a very specific case to mislead by the ridiculous, obscuring the more important point that all randomizations will be as unlikely beforehand as the Shakespeare myth (were as many letters to be typed). It is this that reveals more of the powerful advantage of the randomization device in statistical design. Instead, the monkey story gives the impression that glorious coincidences "could" occur in which a real statistical design "could" fail for this reason. It won't.

A better way to teach the monkey myth would be to point out the unlikelihood of typing Shakespeare in this way and that all randomizations would be as unlikely beforehand.

Soberingly, experience finds it quite common that someone loses her test procedure on the way to work having taken it home the night before to learn, and needs a replacement before the next shift. So it's best to wring hands over what *could* go wrong and manage to prevent all those. The notion that randomization "could" fail is as likely as a monkey typing 1,2,3,4,5,6,7,8,9,10.

Once, a case in a call center with a few dozen agents appeared to have "failed" in its randomization. An intervention that had been left blank (i.e., nothing was tested there) had a highly significant result. The naive conclusion would have been that doing nothing really helped a lot. On inspection, it was found from the researcher's original notes that the first draft of the design had set up Spanish versus English-speaking agents in support of a voice-response intervention. At some point that intervention had been dropped and forgotten but the constrained randomization left intact. So the real finding was that operating in one language differed from the other in performance. This had not been known before. A better design (inasmuch as the objective was to improve performance regardless of language) would have been to randomize completely. This would have been clearer in providing findings over and above the language effect. The design still worked but would have been optimistic on significance by artificially controlling for the language effect.

8.5 PROOF ISN'T IN THE PUDDING

It might appear the proof of the statistical design is in the implementation pudding. Instead it was in the statistical design. If the pudding isn't quite right in the eating, it's recoverable fast, as cases have already indicated (and how). Of course the implementation needs to "work" because that's the only objective.

In seeking pudding proof, it might also appear tempting to run a refining design (to re-prove and perhaps also fine-tune the findings) and a randomized control trial after that to prove the implementation has worked. This is akin to sending out a rowboat to see if it was safe to sail the ocean liner that just passed through.* Instead, the energy is better, faster, and more thoroughly spent on implementing the main statistical design. This is not really an issue of having "good implementation controls." It's more one of knowing what's going on and correcting in real-time using statistical control.

The second and third tests in the last paragraph would have the intent of establishing more "confidence" that the findings were real and would implement. It would look for repeatability of the findings. This would miss a central advantage of statistical design.

Statistical design establishes a wider inductive basis [1] that trumps simplistic repeatability (in which if something works twice or thrice it's reassuring we're on the right track). That wider inductive basis means simply that findings are more likely to generalize. This is designed in because if an intervention works, in the context of so many other interventions also being varied, then it really does work, in the real world (not an artificially controlled one). Breaking down any statistical design in this book, it will be found that each intervention is tested many times, by exactly one half of the total test. The analysis then aggregates all of them to gain the advantage of averaging and establish significance (or not). Therefore in a test involving (say) 20 nurses we're testing each intervention 10 nurses against the other 10. It might appear that a traditional test versus control for a single intervention enjoys the same advantage if the same 20 nurses were in it. In fact it does not. In the single intervention test, all other conditions are held constant (as best we can). In the statistical design, the 18 other interventions deemed important for now are varied in precisely structured ways, defined in the mathematics. Thus the finding will be more robust in

---

* This mangling of a well-known quote on robustness extends to the wider inductive basis of large statistical design.

the real world, meaning it will be more repeatable by design. This feature is strengthened by including environmental variables in the design, which ordinarily might tend to thwart its repeatability.

If a logic were adopted of repeating a test to re-prove the first test's findings and re-proving both sets of findings in implementation versus control, then further testing would also be logical. In other words, if testing twice were really better than once and testing thrice were better than twice, why would not four, five, six times and so on be even better?

In fact statistical control does accomplish this seemingly idealistic position. It's an unforgiving test of implementation and more powerful in view of its investigative properties than retesting. It does continue at a sparing level of effort, for weeks, months, sometimes years to sustain improvement.

Mathematicians use a form of proof called induction in which it's proved that (say) an equation works for the smallest case and also for a typical case and for the next largest case up from that. Then all that's extrapolated so the equation works on anything (i.e., in general). The real world isn't so simple. There's not an equation that always works. Therefore statistical control remains suspicious of something breaking down in implementation and catches it immediately, making it easy to fix in real-time. In other words, we're always suspicious that the particular conditions under which the testing was run (although every effort is made to make sure they're not special) were not general. A pragmatic approach is then to adjust as we go along.

Therefore the scientific meaning of *induction* (being to generalize specific test findings by implementing them) is not as simple as the mathematician's meaning of induction. Mathematics is important but does not fully define the real world, especially as it keeps changing a little over time and some other things.

8.6 WHAT SCIENCE LIES BENEATH IMPLEMENTATION BEING THE HARDEST PART?

When findings from a statistical design are implemented, the improvement looks more like flood waters rising than a tsunami. There are exceptions where it works overnight and stays, then everyone celebrates, surprised that they pulled it off. When those flood waters are rising, in the first day (or week or month depending on how frequently the improvement is monitored) opinions will differ on whether the implementation is working. Experience with statistical control, using correct theory, allows a series of early signs (or warnings) as clues to reinforce (or correct) the emerging improvement.

An analogy lies in planting seeds. Everyone knows seeds cause plants and flowers after awhile (and some other things such as water). Early signs of success are green shoots and later stalks and leaves, then finally a flower. Early signs of success also happen out of our sight, especially properly forming roots. If someone knew only a little about how seeds cause plants they might think to pull the plant up and take a look to see if the roots were forming correctly. This would likely kill it. If the seed planter later argued seeds caused flowers the root inspector would tend to argue that perhaps they work elsewhere but not in his or her particular garden.

When implementing a statistical design, the early clues (or warnings) are subtle in reading the statistical control chart. If they look one way then everything is as it should be and the implementation can be reinforced as is. If they look certain other ways then immediate action and expertise are needed. The cases indicate by example but every case differs from the last.

The early clues are not obvious to the untrained eye and thus need some help. This needs the manager/scientist collaboration. As was so clear in the opening retail case, the two specialists collaborated. Had they not, natural forces from reasonable people working in the trenches would have taken down the emerging improvement. The same was so in the opening chapter's clinical care case. These and the other cases as well as case fragments in the main text give a rather clear picture of how improvement and innovation really work and how it looks before it looks good.

What makes implementation so hard without statistical control is that everything naturally varies all the time. Statistical control theory shows that we will not know the reasons for most of these fluctuations. People learn deterministically: meaning if we do this we get that. So reasonable people think if we implement this we should get the 15% improvement the statistical design predicted, right now. In all the cases provided, the improvements took a week to three months to be forced to mature. They would have remained elusive without scientific forcing behind management skill. After that work, everyone could see, deterministically, that they implemented and got improvement, with hindsight. Usually they did not believe it was on the right track in the early days. This is because there was little to grab onto at first, deterministically.

The real world is stochastic, however, meaning if we do this then we get that give or take a little (or a lot). Getting "that" often also depends on several other things. Statistical control reveals these as they happen. Even if those several other things happily line up for us (exactly as they did in the test phase) then the stochastic world we live in means that when we keep doing "this," we get one or two steps forward and one back all the time. It turns out that process improvement happens suddenly after a little teeth-gnashing. It causes a little surprise even though it was on its way for awhile.

Reading the early clues in statistical control charts has a skill to it (even though it's self-teaching) and to the untrained eye it just looks like a series of dots. The larger the scale of the problem, the harder this aspect becomes as the floodwaters tend to rise more slowly. Few processes have been seen to be slower than a month or two to implement. Some statistical designs do take a year or two to run but they are rare and implement much more quickly. Most statistical designs are done in one to three months, implemented in less time.

All processes, even the large-scale ones, work in the same way. They all follow the same rules to be exploited then tamed by statistical design and control. In statistical design, using the randomization device, the real stochastic world is translated to the deterministic one from which people can learn clearly. Therefore, in a statistical design, we did that and got this. At implementation, as essential and helpful as statistical control is, we're forced back into the stochastic world in which we live. As with planting seeds, statistical designs invariably work but need a little nurturing at implementation. Implementation does not spring up overnight, usually.

When advising that an implementation is (or is not) on track, the scientist puts her credibility on the line. Over time everyone starts to see how it works but that takes a case or two. Proper theory and experience is therefore important when leading projects technically.

A little knowledge of physics ought to allow anyone to place a heavy bowling ball on a rope attached to a high ceiling at a 45° angle to the vertical and release it without any additional push, from a position just touching the forehead. It's known the bowling ball will swing all the way to the opposite side of the room and then swing back stopping again a little short of the forehead. Few of us will try that, however. Process improvement requires trusting the science and the mathematics then standing behind the statistical design's findings using statistical control properly. It also requires immediate action if any of a number of issues occur that can thwart the improvement (akin to the rope breaking as the bowling ball approaches the forehead). Statistical design and control are unforgiving skilled work that everyone contributes to, causing high return for the enterprise.

Statistical design results are easy for everyone to read. Statistical control isn't, except with hindsight. The first appears deterministic; the second remains stochastic. A similar experience everyone has is trying to decide the bottom of a recession when it's happening. Statistical design makes the problem appear simple. It lists 20 or so things and what they do to the process so it can then be improved. It is propelled by the randomization device not seen in real life. Statistical control operates in the real world again thus it is harder at first, but not hard for long. The single test that everything was done properly is that something improves. If we miscalculate, design poorly, overshoot our headlights, or try to put a better foot forward than the one we have, the process will refuse to improve. Nature won't be fooled.

8.7 COMMON IMPROVEMENT STRATEGIES

Improvement strategies (e.g., PDC(S)A: plan–do–check(study)–act, DMAIC: define–measure–analyze–implement–control) all use the scientific method. The repeating induction–deduction cycle used in the scientific method is direct. It's really done intuitively (as the kid with the balloon demonstrated) and no one really stops to think "induction–deduction" all the while.

Specifically, statistical design and control both mirror PDC(S)A. DMAIC fits statistical design into the "analyze" phase and statistical control traverses the "implement and control" phases (except "control" is started and continued with "define"). Within Six Sigma there is a different procedure from DMAIC for designing new processes. With statistical design and control the same procedure is universal. Pseudo-improvement strategies that exclude statistical control will be less effective. PDC(S)A and DMAIC both feature statistical control at their centers.

Improvement and innovation are possible without statistical design and control. There's certainly more than one way to skin a cat. It is, however, efficient and powerful. It should not be used on everything at once but on major problems especially where other efforts have been unsuccessful.

8.8 RANDOMIZED CONTROL TRIALS (RCT)

Randomized control trials (RCT) remain important, although less pure than orthogonal design. It depends on the problem: if the need is to prove out a treatment, RCT will do so and has the advantage after 60+ years of being a common language. If the need is to solve a complex problem, a statistical design and control offers a powerful way, such as to optimize a treatment at the same time as proving it, by varying dosages, frequencies, concomitant treatments, and so on.

The protection offered by blinding in human subject testing is enhanced in large orthogonal designs in that no subject can possibly know which interventions are helping. Only the researcher analyzing the data can see that. The test is also made more real world in providing every subject with a set of treatments rather than the knowledge that they have a 50–50 chance of getting the control as a sugar pill or placebo.

RCT is but one way to ensure homogeneity among test units. Statistical control offers another way that applies in general, especially where individuals cannot be randomized.

Researchers may find concern in that the counterfactual in an orthogonal design to test (say) 20+ interventions is not the same as a strict control. Instead the counterfactual indeed has the intervention untested but all other interventions will be found in exactly one half of that counterfactual (vs. not in its other half). So the argument tends to form that orthogonal designs do not use strict "controls." Although that is true, the real issue is which more faithfully follows the real world. The test versus control appears more pure because it is in more common usage today, but in fact orthogonal design is pure real world. Both approaches remain helpful.

The false alarm rate for large orthogonal designs was explained earlier showing its rate reduces. It's straightforward to adjust to realign the proportion of findings by chance to 5% but completing the scientific work is suggested instead.

8.9 STATISTICAL DESIGN AND CONTROL ARE FOR REAL PROBLEMS WITH EVERYONE CONTRIBUTING

Statistical design and control need expert technical support and leading. Unfortunately it's one of a few fields where it can't be learned by study alone; the mathematics quickly falls short of the real world. The easiest way is to make a start and fill in the theory as it's encountered. It's not possible to become fully knowledgeable by the mathematics alone. This never stops, as what's needed for the next design is rarely covered by all the previous ones or the textbooks.

Fisher's writing on this with additional comment [5] remains important:

> ... design experiments ... in the real world and answer for the consequences of their advice, [not] how optimal experiments might be designed under ... hypothetical conditions ...
>
> Teaching mathematical statistics in this way ... led students to think not of the reality but of mathematics as holding the key to statistical understanding.

Talks at junior high school through university statistics faculties (using different language and depth) have found similar interest in the subject upon hearing it, usually for the first time. This suggests it's not hard, but stimulating enough for such a wide range of experience.

What has captured people's attention in industry and research has been the catalysis of imagination (in dreaming up interventions) and the thrill of cracking a long-standing problem. The financial results follow. A few times a year, meetings have erupted into cheering, applause, and so on, as a troublesome problem has been cracked. There has been a moment when each puzzle is completely solved and everyone knows it at the same instant. This has usually been at the implementation breakthrough, as the control chart arrives after a few days or weeks of people not quite knowing if the experimental findings really would work in real life. (Because the experiment was real life too, it will work; not all know that when embarking on the project.)

Statistical design and control has been found both to awaken, and be caused by, curiosity, like the kid with the balloon. In their own way, experimental solutions by working groups may be as satisfying as to the scientist who held an apple (as moon) up to the moon (as earth) and realized how the whole thing worked. Most problems are much easier, however, and organizations generally have been found capable of solving them.

9

Managing Improvement and Innovation

The long-range contribution of statistics depends not so much upon getting a lot of highly trained statisticians into industry as it does on creating
a statistically minded generation...

Walter Shewhart*

9.1 ORGANIZATION

Improvement starts at the top and stays there but the overall effort is not
top-down, as the cases have shown. Most cases in this book were initiated at
the C-level, translated nontechnically. Typical time needed at that level has
been five minutes a week plus occasional meetings to review a completed
case for action. Regarding starting at the top and staying there: a monthly
review in an hour or less of the statistical control charts sufficiently manages
increasing competitiveness. The items reviewed together follow the company's strategy. Each chart either is (or drives) one of the strategic objectives,
such as market share with a couple of sales statistical designs underway or
complete. The completed efforts stay under review for sustaining and for
continuous competitive progress. Executives are typically an intelligent and
curious cohort and this has been found useful to the scientific work.

Equally important are all levels and functions. Statistical design and control are again translated into everyone's language. Much of the scientific
work is done in the trenches, at the front line. It's especially important to
learn from the craftspeople, whatever that craft is. Nurses, mechanics, reps
in a call center, engineers, doctors, tree-cutters, electricians, actuaries: they

---

* Shewhart, W.A. (1939, 1986). *Statistical Method from the Viewpoint of Quality Control*. New York:
Dover.

all know about how the thing works and something about how to make it work better. Organizations have been found to follow a rule of thumb of about 40% adopters, 50% neutral, and 10% against statistical improvement/innovation. That's more than enough support to manage competitive improvement and the 10% against are critical to challenging the work properly and making sure it is safe for the customers and the market.

9.2 SPEED WITHOUT NET RESOURCES

The statistics move faster than any organization can. When the next step in improvement work comes, the statistics will be there, waiting. Managing competitiveness has as its most vital elements speed and minimal or no net resources. It tends to take a year or two to move from minimal to no net resources. One statistician (already on staff within most organizations) can support 10–20 projects at a time, which is about as much as the organization can digest.

There's no exact answer on speed but one to three months rarely needs to be exceeded by statistical design, plus typically less time to deliver and sustain the improvement with statistical control. Often a later time will be preferred, except when it comes. If the need is now, it can be worked into the live business. At the same time as the product or service is being made, the same data used to manage can be used to improve or innovate and is free. In the modern competitive world it's wasteful to use the data only to produce. They can also be used to produce more profitably.

9.3 HOW TO MANAGE SPECIFIC IMPROVEMENTS/INNOVATIONS

The technical work doesn't follow a set list of steps but the following tasks always occur and make specific projects easy to manage.

1. Define objective(s), measurements, and goals. More than one objective works fine with interventions aimed at each. Sometimes interventions aimed at one measurement surprisingly help another.
2. Specify baseline process data and run a quick measurement error study.

3. Discovery, including research sessions with a cross-section of people and levels, to list interventions, knowing the thinking and the physics behind each.

4. Analyze baseline and measurement error data to give:
   - Process control chart and a way to keep it maintained
   - Measurement error precision, stability, and whether an acceptable percentage of total process variation
   - Measurement control chart, perhaps integrated into process control chart hybrid
   - Homogeneity, or lack thereof, among test units (if they're running at the same time)
   - Stability among test units if they're temporal (e.g., consecutive manufacturing batches)
   - Calculate test precision, sample size, duration with a pinch of salt
   - Complete work to bring measurement error stability and homogeneity among test units if needed

5. Prepare the rough statistical design.

6. Develop each intervention into a standard procedure (vs. typically status quo as counterfactual).

7. Integrate #6 into #5 and complete the statistical design resolving any clashes among pairs of interventions.

8. Develop a light-touch adherence plan with an intent-to-treat approach.

9. Educate everyone who'll have a hand in running the test.

10. Run the test, checking its data are flowing and captured correctly and using management controls to correct any important adherence shortfalls. Deliberately stay with the test and the people running it to know what goes on at all times and places as it unfolds. Check the control charts for insights and likely improvement during the test.

11. Analyze results, calculate improvements, rough out the draft implementation plan.

12. Don't present the results yet!

13. Discovery to complete an understanding of what went on in the test and explain, with help from the people who ran it, why the findings happened. Stay on elusive findings to understand why, in a few hours or days, with extra peripheral data and records.

14. Discuss results and findings and complete the implementation plan, owned by the department(s) that ran the test.

15. Assign the person(s) who will be responsible for the improvement.

16. Implement while spotlighting the statistical control charts (process and measurement error) and help managers use as a simple feedback-control system.
17. Close any gaps and shortfalls in implementation by monitoring adherence lightly, and using the statistical control chart signals, work out what's happening and how to fix it.
18. Sustain long term using statistical control and improve more later if or as needed.
19. Calculate financial return and verify with finance.
20. Use the case as insight for other project personnel.

9.4 STATISTICAL DESIGN AND CONTROL SUMMARY

The advantages of statistical design and control, using only techniques available in the literature, are self-evident. The basic idea is simple. The main disadvantage is that it requires some technical skill and management support. Those might also be argued as advantages over the competition.

The essential elements in application include establishing homogeneity and stability in both the process and the measurement error, although there are workarounds if unattainable. Statistical control isolates then exploits uncontrolled variation, keeping an eye on the economic aspect. Developing strong interventions and correct randomization are central to statistical design. Implementation that would ordinarily be difficult uses statistical control again, as a simple, real-time, feedback-control system.

Randomization has surprising properties shown here by case examples and simple exercises on data. Its facility to isolate solutions from large amounts of clutter can be applied in any business problem. Used in orthogonal design it finds cause and effect that would otherwise be more difficult or controversial. Intent to treat, leaving open an element of laissez-faire, was stressed in real-world testing to give results that carry long term.

In the cases provided, about a fifth of interventions helped and 5–10% hurt. These ratios include expert ideas tested as interventions. Not much correlation was found in the examples between prior expert opinion and actual findings. However, the way to find strong interventions is through the broadest expertise within the firm and sometimes with its customers, vendors, outside experts, and so on.

Many of the cases find a reduction in variation during the statistical design and a degree of improvement. This is largely due to the increased standardization. There's no mathematics to predict or explain this, which is central to statistical control. Statistical design and control have been found most effective on a fairly large scale and make most sense for larger or complex problems, especially if unsolved by other means.

It's not possible to substitute an analysis of historical data (although those are often effective too) because the business is not stressed in new controlled ways not tried before. Also interactions are difficult or impossible to analyze reliably in happenstance data. Historical analysis listens to what the process has said before; statistical design and control interrogate it rapidly then keep interrogating it through statistical implementation control.

The case was made for single large statistical designs backed up by dynamic statistical control. Additional testing is fine technically if there's time and patience. To make something improve suddenly and hold that gain using investigative statistical control has been found efficient and likely to optimize the process in its present form. As a strategy it entertains both the possibility that a single solution set stays rigid over time, and also that the solutions are shifting over time in ways unpredictable but diagnosed by the rules of control.

The creativity behind the interventions needs work. It's not just the quantity tested that matters; it's the ingenuity behind them.

Both statistical design and control were designed for the real world with everyone's input and use. An effective way to get started is just to copy one of the cases, making adjustments for its differences to what's found. The designs are always the same and much is learned on the first case. If staying close to a completed case, nothing more than simple arithmetic is needed. Repeated uses of the same design on several different problems keeps revealing more about how they work. A risk to avoid is testing too few things, especially because that may appear safer. Its economics tend to bring less competitive advantage.

All designs go wrong in one way or another; usually several small things don't go according to plan. Being ready for this and fixing it immediately is usually common sense. An unwatched design wouldn't work. It needs deliberate interference by whoever assembled it and plans to analyze it later. It belongs to whoever runs the operation.

The single test is that something improves suddenly and sustains, using rigid rules of statistical design and control. The competitive advantage from statistical design and control comes from randomized orthogonal designs that establish cause and effect directly, and the scientific method, which is just a way to understand how things work and make them work better.

While the mathematics is well defined and published, adoption (of especially large statistical design) in industry remains limited. Talks and discussions with researchers and academics have often brought the seemingly rhetorical question: "So this is just experimental design then?" in much the same way as one might have asked in 1960 of a plan to put someone on the moon: "So this is just physics then?" In responding, it has been helpful to ask first whether such designs have been run to improve something. The answer has most usually been along the lines of: "No, but..." as if one easily could have but hadn't yet (and as if study of the mathematics alone were sufficient).

It is into that trailing "but..." that this book inserts itself.

Appendix: Answers
to Exercises

1. It's the difference between test and counterfactual that matters. That difference should be bold enough to overwhelm any lack of standardization. The standardization of the test condition follows the intent-to-treat argument: that it not be better standardized for the test than can be accomplished long term.

2. Randomization ensures each intervention effect is measured independently of other influences, known or unknown, large or small.

3. With about 20% of interventions found empirically to help, the false alarm at 2.5% (on the helpful side) is dwarfed. In a statistical design with 20-ish interventions, about four would be expected to help. A false alarm on the helpful side would be expected at zero or one. In a small design, the same naive approach finds no such reassurance. The worst that can happen in the example scenario is only about three of the (say) four really help, in which case the fourth is unlikely to flip all the way to hurting. So the larger the design, the lower the relative false alarms. This relies on the empirical finding that roughly 20% of interventions help in large designs. If that number were closer to 2.5% the large designs would not work well.

4. Intent to treat allows complete freedom, emphasizing the word "intent." It's not cavalier or capricious: the intent is that the intervention be tested. Thus within reason it's managed rather tightly in the statistical design. If an intervention turns out to have been tested at about 50% strength (whether the percentage is known or not) then whatever the result is, the same one will follow at implementation. Were a way to be found that increased the 50% closer to 100% then the payoff would be expected to roughly double. Intent to not treat is found equally useful. The counterfactual is supposed to not get the intervention. In practice they often have a slight semblance of it, so the test really becomes a reinforcement of that, to find out if reinforcing helps. Therefore it's the reinforcement of the thing (rather than the thing itself) that's being tested. It's ill-advised to try for a pure test of the thing versus its complete absence. Presumably the thing

is controversial (or at least it's unsure or not credible as to its worth) or it would already be done most of the time. The "pure" test would draw the reaction from both proponents and detractors of, "Of course!" and the test would move the collective no farther forward. In a recent case, typical of dozens, an intervention was so popular as word spread on the grapevine, clandestine adoption by the counterfactual started slowly then spread like a virus. In the weekly interim analyses it could be seen starting as a large effect and then diminishing to a fraction of full strength. This did not of course mean that its effectiveness was dropping, rather that the counterfactual was catching up to the test group. None of this detracts from the well-managed statistical design. At its end the adherence to the intervention was tallied roughly and found to be at least three quarters in the counterfactual. With a couple of points of improvement at face value, the real effect would be as much as four times what was measured. It would be folly to assert a more aggressively managed design that kept everyone rigidly in the test or the counterfactual. Nature doesn't work that way. The whole idea was to find things that help, for the first time, innovatively. Here's one that the science and the people find attractive and no one's even having to ask them to do it, as they flock to it. It would have been a mistake to enforce, rigidly, the original plan. If indeed that could even be done. Therefore a percentage of adherence is perfectly fine at whatever it is, provided the intent were managed. Real examples go through the entire spectrum from near perfect adherence to minimal. It has to be known what went on behind the adherence. The argument here would not change whether the intervention results were significant or not. They would only be noticed if significant, however. The whole idea is to test things, but to test those things as managed in the real world, often not as first expected. If someone felt such tests ought to have been more strictly managed, one could only wish them luck with that. Also the findings would be less useful, if they could be found. The "–30%" adherence part of the exercise follows by extrapolation of the above, or by Answer 18 below, explicitly.

5. False alarms (whatever those really mean) are known with certainty when something improves or doesn't. If a statistical design stopped with publishing its results, the full improvement is unlikely. Significance (or lack thereof) is followed by the rest of the science and statistical control. No harm can accrue inasmuch as a false alarm

intervention is unlikely to really hurt. In the case being entertained by the text when this exercise was set, an updated analysis was run with an additional couple of months' data. The story was unchanged other than P grew to most significant. This gave stronger proof.

6. In several of the cases presented here, regression models were run to "remove" the effects of other variables (e.g., HCC score, regression-to-the-mean, prior trends, etc.) without changing the story other than finding usually more significance. The weakness is that the models can be invalid and miss what randomization is doing.

7. As noted later in the text, additional testing or an implementation control presume a correct solution that remains fixed. Statistical control provides a more dynamic way to make an approximate solution from a single test work, adjusting along the way for conditions not experienced in the test. No harm will accrue from refining or validation testing if the organization has the patience for it. The cycle time for innovation would be slowed. Foldover designs (actually what are called Resolution IV designs or better) are preferable to keep pair interactions clear but this is often not possible if the design is to be large enough to make a difference. Taking the logic of repeated testing literally (that three tests are better than two which are better than one) then the idea of statistical control is to take that further into a continuous fine-tuning.

8. The analysis of the retail case was given in the text. Manual calculation in the first few cases experienced gives a better understanding of how the designs work. Regression is usually used in some form, with the test versus counterfactual programmed as 1 versus 0 (rather than the +1 vs. −1 suggested by the designs).

9. The split line on the otherwise straight center of a probability plot of intervention effects means one row has an unusually high or low result. In this case it traces to row #15. Both stores have very close results, therefore it is reinforced as real. It's not a problem because there's plenty of improvement in the findings already. Some instances can be useful to further improvements. The normal plot shows this particular split has no effect on the significant findings. The way to trace which datum is extreme is by noticing it would increase about half of the effects and decrease the other half. The normal plot identifies which effects (within the split straight line portion) are high and which are low. This will be found to tally fairly closely with one of the row's ± signs. That is shown in Chapter 5 and traced to a single row. This is an extremely useful tool that is difficult to match analytically.

It also clarifies that A, AG and two more interventions are significant in that the parallel lines extended indicate significance (rather than a rough line through them both).

10. The additional tests for control help with not missing slight changes in process data that do not break limits.

11. Pencil and paper and a few minutes with the data would work well. The experienced ear would suspect immediately no useful cause of all three plants dropping at once. With the data plotted on three control charts, the answer is confirmed. It's colloquial but pragmatic to advise that the three dropped by chance like three coins landing heads. The real answer is that the many complex causes would be uneconomic to pursue under statistical control theory.

12. The economic loss from pursuing the three plants dropping, as if there were usable subtending information, would be in using resources essentially on a wild goose chase. The opportunity cost is that the resources could be used to improve all three plants so that capacity were higher. Chapter 6 gives the example. Of course all three plants will still drop together about once every eight months. They will all three increase with about the same frequency. It's better, however, if they're doing that from a higher base. (There are six other patterns of one or two plants up or down. Only the "all 3s" are noticed typically and thought to be problematic (if down) or good news (if up).

13. This one was answered in the text. Recapping: the reason removal of measurement errors (as symptoms) when in statistical control is that their underlying reasons will not be prevented. Fixing at the source only the uncontrolled measurement error makes the measurement usable and stable. If the symptoms are removed for specific data pulls, it tends to throw the measurement error back out of statistical control, for example, with the symptomatic errors that are missed. This is a little like a logger removing branches on a felled tree so that it can be transported easily to the sawmill. No attempt is made to prepare a "perfect" trunk. That's the sawmill's job. Any attempt to use a chainsaw to make the trunk look smoother would be clumsy and do more harm than good.

14. The partial replicates appear unreasonably close to the original runs of each row in the test. Also, having just four degrees of freedom is stretching the small sample statistic a little far. Still it's the best information available. Had there been time, full replication would have answered this. A couple of things were not known by the study

people. One was how closely pulses would be expected to replicate. It may have been that pulse rate is quite precise in this way, given the conditions set up. Another was the extent to which subjects could influence their pulse. A flaw in the measurement of the partial replicates was that the subjects were looking at the original result and knew what it was as they took the second (replicate) pulse. No one knows if subjects would be biased even subliminally by this knowledge. In the real case, it was written up that there was suspicion of the natural variation being underestimated. Pragmatically, no change in the story occurs if so. Obscuring the whole study was the unnatural loading of the test with interventions known to affect pulse. There was the heavy backpack surprise but the interventions were more or less designed to make a difference. A better design would have been a fractional factorial in 16 runs with 8 interventions. Those can be worked out from scratch in a few minutes. Better yet would have been a multifactorial in 12 runs with 11 interventions but no one could construct those from memory.

15. The analysis suggested is unnecessary because randomization eliminates the sales manager skill from affecting the design's results. If the sales manager's name (which modern software will read) is built into a mathematical model then a couple of problems occur and the mathematics is overdriven. Something called multicolinearity is likely (where complex dependencies occur in the input data) and with only 31 degrees of freedom in the design for analysis of 8 interventions, 7 aliased sets of pair interactions, and 19 sales managers, it would be asking a lot. Also, it misses the point that this would not mirror what the staff sales managers did. They did not look at just names of their sales managers when balancing each intervention with its counterfactual in regard to performance. They looked at the performances they knew from experience and weighed them all intuitively. If a performance number were used instead of each sales manager's name we might be in business (and there'd be enough degrees of freedom). But it's all unnecessary as randomization trumps and prevents all problems. The model is likely to err whereas randomization will not. The point though is a simple pragmatic one. Given that randomization would do its job, the issue was to make sure managers saw how. The approximate exercise allowed them to. One real issue that could have occurred would be if there were (say) only three sales managers and one had far stronger performance.

That then requires a different design in which the sales managers are blocked out, perhaps by building them in as pseudo-interventions. Randomization and blocking are two ends of a similar intent in good design. Another option with the alleged three sales managers would be to stratify them so that (especially) the strong performer ran all 16 rows of the design. If an intervention could form an interaction with sales manager performance (e.g., stronger performers did especially well with given intervention) then stratification is preferred.

16. The two stores have highest sales volumes. Similar phenomena occur quite often under correct randomization.

17. A practical trick is to include the pretest admit rate in a regression model as a covariate. There are flaws with this but it typically confirms the main analysis. In real care management cases, this has sometimes been found significant, other times not. Either way it removes regression-to-the-mean from suspicion of "causing" the results. The cases can be run through this procedure and will be found reinforced. This is just a special case of confirming that any variable did not influence the results in a randomized orthogonal design. It's known at the start under randomization but the question often comes up. It's put differently in different industries. However, it's always the same fundamental question.

18. One real intervention that appeared to hurt was in fact due to 76% adherence in the test and 94% in the counterfactual. That's a difference of 18 percentage points. Thus in fact the intervention helped even with just an 18% point discrimination. The real effect would therefore be larger than the measured one (naively: about five times, but of course that didn't happen as large in practice). This was a student intervention and the test involved requiring them to attend an extra session. Therefore the other finding was that kids wouldn't attend as much if told to but if they heard about it they'd clamor. The following semester just put the word out about the helpful session. The test was well managed and revealed that the session helped as well as showed how to increase attendance. It is easy to see how this happened in the real world which includes human nature.

19. a. Programming modules form the test units for interventions in design through testing and into user acceptance. Each "bug" traces back to the module. The usual view that each module is different yields to the homogeneity check. Software engineers

and newspaper editors were found to have more trouble accepting that their skilled creativity wasn't discriminated but this is no different than all the other case types.

b. Web/SEO applications of statistical design are by now common. The test units are often time windows, taking into account the design traffic patterns over time. The testing of large quantities of interventions is far easier with typical volumes.

c. Trees and squirrels cause power outages by growing into power lines or chewing them. The test unit is geographic grids with interventions around how trees are trimmed, the equipment used, and consultations with residents/businesses. Users want both trees and power, therefore they're in tension.

d. Examples of artistic layout interventions appeared in Chapter 6. It's useful to find a real measurement (e.g., retention rate) to verify later the conjoint.

e. Surrogate measurements (usually near-misses or backup systems triggering) are tested in the statistical design. The usual tendency for improvement during the test mitigates critical event risk.

f. Combining the retail store case, (b) and (d), is common. It's intriguing in artistic layout designs of any description that they can be seen. It's near impossible to really see the 20+ interventions shifting around even though they are all in plain sight. Most designs are more invisible with only one view of one row in view.

g. The designs set up for this case had about 100 interventions in a set of five simultaneous designs. The trick was to tag complaint data to all reps of all types who had touched the eventual complaint over its history. Data systems captured easily. The constraint that each design contain each other was met inasmuch as any rep in any call center could have touched any complaint. (Only a few did.) The flows were haphazard, mitigated completely by randomizing all reps to all designs.

h. This problem type (sometimes called "repeat calls") works identically with the call center sales cases but measures a closed cohort captured at the first calls. Then it tracks which have repeat calls inside 30 days. The assumption is that the callback is for the same reason but even if there are new calls for new reasons, their bias will randomize out of the interventions.

i. See (h).

 j. Test units are the sets of three crews and their CO. It's a single design.

 k. This one depends on how the thing works. If sales by all channels are to geographic areas centered on the stores then it's a single design with interventions in all channels at once. If the website is generic, nationwide, then two simultaneous designs will be appropriate. Each run of the website design (say for a day) contains the entire store design.

 l. This combines the features of any online design (e.g., test units per short time window or by geography) with tagging transactions to call-ins.

References

1. Fisher, R.A. (1935, 1971, 2003). *The Design of Experiments*. Reprinted, 2003. Oxford, UK: Oxford University Press.
2. Shewhart, W.A. (1931, 1980). *Economic Control of Quality of Manufactured Product*. Reprinted 1980. Milwaukee, WI: American Society for Quality Control.
3. Plackett, R.L. and Burman, J.P. (1946). Design of optimum multifactorial experiments. *Biometrika*, 33, 305–325 and 328–332.
4. Fisher, R.A. (1926). The arrangement of field experiments. *J. Min. Agric. G. Br.*, 33: 503–513.
5. Box, J.F. (1978). *R.A. Fisher: The Life of a Scientist*. New York: Wiley.
6. Wu, C.F.J. and Hamada, M.S. (2009). *Experiments: Planning, Analysis and Optimization*. Hoboken, NJ: Wiley.
7. Daniel, C. (1959). Use of half-normal plots in interpreting factorial two-level experiments. *Technometrics*, 1: 311–341.
8. Box, G.E.P., Hunter, W.G., and Hunter, J.S. (1978). *Statistics for Experimenters*. New York: Wiley.
9. Grant, E.L. and Leavenworth, R.S. (1988). *Statistical Quality Control*. New York: McGraw Hill.
10. Shewhart, W.A. (1939, 1986). *Statistical Method from the Viewpoint of Quality Control*. New York: Dover.
11. Deming, W.E. (1960). *Sample Design in Business Research*. New York: Wiley.
12. Moen, R.D., Nolan, T.W., and Provost, L.P. (1999). *Quality Improvement Through Planned Experimentation*. 2nd edn. New York: McGraw-Hill.
13. Box, G.E.P. (1966). A simple system of evolutionary operation subject to empirical feedback. *Technometrics*, 8: 1, 19–26.

Index

Note: Page numbers ending in "e" refer to equations. Page numbers ending in "f" refer to figures. Page numbers ending in "t" refer to tables.

About the Author

Kieron Dey studied mathematics and statistics at Reading University, England and management at Rensselaer Polytechnic Institute, New York. He was on the experimental staff at Hirst Research Center, London, England (an early specialized center for applied scientific research), and apprenticed with Joan Keen, a pioneer in industrial statistics. He later joined IIT Research Institute, another contract research organization, serving in several roles including as a scientific advisor. He has held technical leader positions in corporations up to $2 billion in size, now with Nobigroup Inc. He has vast experience with corporate and government leaders. Dey is a Fellow of the Royal Statistical Society.

Printed and bound by CPI Group (UK) Ltd, Croydon, CR0 4YY

23/10/2024

01777697-0003